SolderSmoke

Global Adventures in Wireless Electronics
From Crystal Radios to Podcast MP3s
With Occasional Forays into Astronomy, Kites, and Model Rockets

By Bill Meara

SolderSmoke

Global Adventures in Wireless Electronics
From Crystal Radios to Podcast MP3s
With Occasional Forays into Astronomy, Kites, and Model Rockets

By Bill Meara

HBR PRESS

Copyright © 2009 by William R. Meara

All rights reserved.

No part of this book may be reproduced in any form or by any electronic or mechanical means including information storage and retrieval systems, without permission in writing from the author. The only exception is by a reviewer, who may quote short excerpts in a review.

William R. Meara

Visit our website at http://soldersmoke.blogspot.com

First Printing: 2009   This revision was completed in May 2010

ISBN: 978-0-578-05312-7

Dedicated to

Billy and Maria

Possible future winners of the Elser-Mathes Cup

CONTENTS

Preface..1

Chapter 1 Electrically Inclined – Tales of an Electromagnetic Youth................5
    Electrons and Electricity
    Radio Waves
    Some Basic Equations
    Einstein in the Transformer
    Semiconductors

Chapter 2 Off-the-Air—Amateur Radio Goes Into Hibernation................40
    Junctions and Diodes

Chapter 3 Tropical Rebirth—Ham Radio in the Dominican Republic............45
    The Lowly Capacitor
    Resonance and Oscillation
    Series and Parallel Tuned Circuits

Chapter 4 Boatanchors in Virginia—Back in the U.S.A.............................81
    Transistor Amplifiers
    Mixers
    Modulation: AM, DSB, SSB

Chapter 5 Mid-Atlantic Outpost—Amateur Radio from the Azores.............116
    Balanced Modulators

Chapter 6 Urban Radio—Solder Smoke in Central London......................157
    Amplifier Loads

Chapter 7 Rome—Secret Radio in the Eternal City................................180
    Feedback in Amplifiers

Chapter 8 Conclusions—A Brotherhood Without Borders.......................191

Acknowledgements..194

Index...195

PREFACE

"Amateurs in radio transmission, as opposed to the great listening public, have for their field of endeavor the whole realm of physics, with all the natural phenomena of electricity, magnetism, light and sound. In thousands of volunteer laboratories and in thousands of individual training schools, they toil nightly—volunteer workers, amateurs: amator, lover; amare, to love—all for the love of the work and the thrill of achievement."

From *200 Meters and Down—The Story of Amateur Radio* by Clinton B. DeSoto, published in 1936 by the Amateur Radio Relay League, Inc., West Hartford, Ct., page 3

It was hot as blazes in Santo Domingo that 1995 night, and the humidity was near maximum. But I was happily ensconced in the air-conditioned radio room of my suburban house. I'm sure my neighbors figured that all the antennas on the roof had something to do with my work at the American Embassy, but they were wrong. The antennas and all the radio gear had absolutely NOTHING to do with the Embassy. In fact, for reasons that I think will become obvious, I'd deliberately kept my off-duty electronics work secret from my diplomatic colleagues. During the day, I was a mild-mannered diplomat for the world's one remaining superpower. But by night, I was a radio fiend.

I sat down at a rudimentary table-cum workbench that I'd built myself. In front of me was a radio transmitter that I'd bought as a teenager some 21 years before. Next to it was a receiver that I'd had even longer. I was intimately familiar with the innards of both, having had to repair them many times over the years. Both predated the commercialization of transistors—both employed ancient vacuum tubes: fire bottles as we called them. In other locations the heat from these tubes had helped keep me warm through cold winters, but in the Dominican Republic it just added to the work load of my long-suffering air conditioner. But I liked the tubes. They threw off a very comforting, familiar smell, and they cast a friendly glow onto the walls of my Caribbean radio shack.

Off in the corner an old Tandy computer that I'd first come across on the Honduran-Nicaraguan border struggled to project a crude map of the Western Hemisphere on the screen. The only thing that was moving was a small cube off the coast of Ecuador. It had a circle around it. The circle showed the portion of the earth's surface that was within visual range of this particular low earth orbit satellite.

I was waiting for RS-12—Radio Sputnik 12. It was a bit of amateur technology that had been grafted onto a large Russian navigation satellite. The amateur radio gear was set up to take a signal coming to it on one frequency, amplify it, and then send it back to earth on another frequency. For reasons that will be clear to those who understand the lure of technical challenges, I found this device to be very interesting, and I wanted to send my signals through it.

As soon as the little circle crossed over Santo Domingo, I began to tune my Drake 2-B receiver. I was looking for RS-12's beacon. Right on cue, there it was—faint, but growing stronger as the satellite rose above the moonlit Caribbean horizon.

Next I tuned a bit away from the beacon frequency and began to hear the voices of other fiends who, like me, had sacrificed a good-night's sleep in order to communicate through this piece of Russian space gear.

Tapping my telegraph key (yes, it did seem very incongruous) I listened with my headphones, hoping to hear my own signal coming down from space. There it was. "Di di di dah. Di di di dah." At this point, I had already accomplished my mission—I'd sent a signal through the spacecraft. But I wanted to do more, I wanted to use that satellite to do what radio amateurs do: I wanted to use the technology to communicate with a fellow enthusiast.

I knew that my Dominican Republic call sign would stir up some interest among U.S. amateurs—they were accustomed to speaking to close-by fellow Americans. I was in a relatively exotic foreign location. RS-12 was too close to the earth to permit any real long-distance communication. 1000 miles was really the outside limit. And within this range I was about as exotic as things got.

I found a clear spot in the cacophony of signals and I put out a CQ, a call to any station that might want to talk to me. The satellite was moving at 17,000 miles per hour—it was circling the earth every 90 minutes, and it was moving so fast that the Doppler shift had to be adjusted for as you used the satellite. I could hear my own voice through the headphones on the satellite's downlink frequency.

As I made my calls, I kept an eye on the computer screen. Time was short—I knew that that little circle around the satellite would move up to the Northeast, and when its perimeter crossed over Santo Domingo, the signals from space would disappear. In the ten minutes it took for the satellite to go from horizon to horizon, I talked to three stations, all of them in the Southeast U.S.

At the appointed moment, the little line passed over Santo Domingo and the radio went silent. RS-12 had gone below my horizon.

I paused for a moment before turning off the old radios and heading to bed. I thought of how strange it was for me, an American guy from New York, to be using these old tube-type radios to talk through a Russian spacecraft from the Dominican Republic. There were certainly a lot easier ways to have communicated with the people I had spoken to—I could have picked up the phone and called any of them. But ham radio has both technical and social elements. Amateurs embrace the technology, but they do so in order to be able to use that technology to socialize with like-minded enthusiasts; the friendly banter in those brief and technologically exotic satellite conversations seemed to capture very well this mixture of the technical and the social.

I went to off to bed that night thinking that my radio work of that evening very much captured the essence of my life as a radio amateur: the old familiar American gear acquired as a teenager being put to use in a novel way from a exotic location. My career as a diplomat required me to move every few years. Each move meant a new job, a new house, a new country, new friends, etc. Amateur radio was one of the things in my life with the deepest roots, one of the few common threads running through it all, one of the things that gave a comforting sense of continuity.

But it was a secret comfort. It seems unfair, but for some reason, there are hobbies and interests that are more socially acceptable than others. Sports, for example. Work colleagues can spend seemingly endless hours discussing the trials and tribulations of their favorite sports teams. They will keep work colleagues updated on their plans for watching the next game, and will later report on their impressions of the event. I think that if I were to engage in an even remotely similar level of discussion of my radio hobby, I'd quickly be tagged as at best eccentric and at worst as a bit of nut case. This may be partly the result of my work—there are just not a lot of techies in the diplomatic corps, and amateur radio would be seen as being particularly unusual in this group. And just imagine the eyebrows that would be raised if I came bounding into work one morning anxious to tell everyone about my late-night chats through the Russian space ship that flies over my house each night. I hope you can see why I've taken a somewhat undercover approach to this hobby.

In August 1966, *QST* magazine published a short article that captures the unique mixture of science, technology, exotic geography, and fraternity that is the essence of amateur radio. The article begins on a hot and sunny day in the year that I was born (1958). A radio amateur named Jorge Barbosa and his little boy were driving through a small village north of Luanda, Angola. Suddenly, they spotted a number of odd-looking buildings and quite a few antennas growing up in the middle of the jungle. They had stumbled upon the home of Carlos Mar Bettencourt Faria.

Bettencourt gave them a tour. He was a newspaper writer with a wide range of interests. He had introduced underwater exploration and spear fishing to Angola, and had written books on radio astronomy and marine biology. His compound had electronic, photographic, and chemical laboratories. Bookshelves filled with technical literature lined the rooms. The antennas that Barbosa and the boy had noticed included a radio telescope for 108 Megacycles.

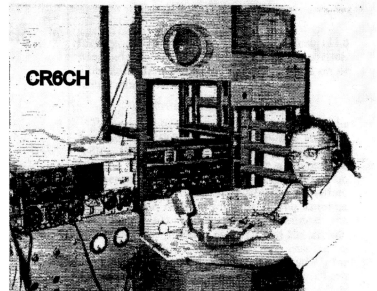

Barbosa told Bettencourt of his interest in amateur radio, and this apparently planted the seeds for yet another hobby for Bettencourt. Soon Bettencourt told Barbosa that he'd obtained his amateur radio license (callsign CR6CH) and had traded his underwater camera for a war-surplus radio transmitter.

Even for an intrepid man like Bettencourt, rural Angola must have been a very challenging environment in which to build radio equipment. When satellites started going into orbit, Bettencourt wanted to build some gear that would allow him to listen as they passed overhead. But he was frustrated by his inability to get some specialized crystals. Barbosa suggested that he make use of the amateur radio fraternity's tradition of mutual assistance. Reaching out from Angola, Bettencourt used Morse code sent through his war surplus transmitter to contact some American radio amateurs. Soon the needed crystals were on the way to Angola.

I didn't come across Barbosa's article until I was in my forties, living in the Azores. (In an odd coincidence, I was at the time living on the small island on which Bettencourt Faria had grown up.) Reading about his efforts to reach out to space from a hot and dusty corner of the third world certainly rang a bell with me—I immediately identified with Bettencourt. He was obviously way ahead of me in technical expertise and scientific knowledge, but I could tell that he was interested in the same kinds of things that I was, that he derived the same pleasure in overcoming technical challenges that I did. To a certain extent I suppose you could say that this guy became a hero of mine – he was building radio telescopes from scratch in rural Angola in 1958! But "hero" is not really the right word—it was more of a fraternal, kindred spirit kind of thing. With a hero you ask for an autograph and stand back in awe. With a brother amateur you can commiserate about technical problems, ask if you can rummage through his junk box, and know that he'll give you anything he can spare. Bettencourt was a Portuguese speaker living in colonial Angola, but after reading about him, without ever having met him, I felt sure that we would have been friends. The title of Barbosa's article was "My Friend, CR6CH." That article about Bettencourt was a reminder that radio amateurs are members of a wonderful worldwide fraternity, a fraternity forged in solder smoke.

This book is the story of my secret life as a radio amateur. I hope to share with you some of the fun and excitement that I've had in this secret life. As in the little Dominican satellite adventure recounted above, much of the fun involves actually using the radios—the book contains many tales of adventure on the airwaves. But a big part of the attraction of the hobby is more technical and cerebral. It involves the quest to understand the science, to understand the physics that permit us to communicate over vast distances. And it involves the development of the technical construction skills that allow us to take that understanding, and mix it with some radio parts and solder to build our own equipment, equipment that will carry our thoughts and voices beyond the seas.

In the course of presenting my secret life in radio, I will from time to time pause and discuss my struggles to understand some elements of radio and electronics theory that I found difficult to grasp. The technical discussions are not intended to serve as an electronics primer. I couldn't write that kind of book; first, there is far too much to cover, and second, I'm not an expert—I don't understand it all. The technical interludes in this book are meant to describe my efforts to understand some of the things in the real technical books (the ones written by real engineers) that left me scratching my liberal arts head.

A glance at my bio may cause you to wonder why you should read a technical book written by someone who holds no degrees in science or engineering. I've often asked myself that same question. For those who are, like me, complete amateurs at radio and electronics, I hope this book will fill in some of the "explanation gaps" that—from my perspective—exist in the radio literature. And for those who do have the letters BSEE after their names (I admit to jealousy here), I hope that the technical discussions might stimulate some thought about what's really happening in the schematics and the equations. I get the impression that there's not a lot of time for this kind of deep "how does it really work" thinking in university EE programs. At the very least I hope the book will let experts better understand how radio theory looks to a layman.

In the book's technical interludes I will share those very satisfying "Eureka!" moments that sometimes come after long periods of reading, pondering, and doodling. I'll also describe the books that I found helpful. I hope that these technical interludes will be illuminating to others interested in these topics. But I must emphasize that no one should expect to come away from a reading of this book with a complete understanding of radio theory. For that you will need many, many books. To paraphrase a controversial politician: "It takes a library…" Over the years I've found that it is very helpful to have at hand a number of different books on radio theory. When trying to understand a difficult topic (and there are many in this field), sometimes you have to put down one book, and pick up another. You may find that one author has a better way of describing the phenomenon or circuit in question. "It takes a library…" I hope this book merits a place on the bookshelves of amateurs who, like me, are trying to understand this very complicated and mysterious subject.

Before we get started, let me ask you for a favor: As you read the book, please keep a pencil close at hand and mark any mistakes or typos you come across. This is, after all, very much a homebrew book, and as in all my homebrew projects I could use some help. Please report any errors to me. My mail address is bill.meara@gmail.com Thanks!

CHAPTER 1
ELECTRICALLY INCLINED
TALES OF AN ELECTROMAGNETIC YOUTH

Mother: (concerned) "I'm worried about little Dilbert. He's not like other kids. Yesterday I left him alone for a minute, and he disassembled the TV, our clock, and the stereo."

Doctor: (deadpan) "It's perfectly normal for kids to take things apart."

Mother: "The part that worries me is that he used the components to build a ham radio set."

Doctor: (suddenly alarmed) "Oh dear... I'm afraid your son has... The Knack."

Mother: "The Knack?"

Doctor: "Yes, The Knack...It's a rare condition characterized by an extreme intuition about all things mechanical and electrical, and... utter social ineptitude."

Mother: "Can he lead a normal life?"

Doctor: "No, I'm afraid not. He'll be an engineer."

Mother: Sobbing

    From Dilbert (video) by Scott Adams

"I'm fairly sure that I was one of the youngest ham radio operators in the country. That was huge for me. But even more importantly, I learned about the process of getting a ham radio license—what I needed to know, what I needed to build the equipment— then I built the radio. It gave me a lot of confidence for doing all kinds of other projects later on."

    Steve Wozniak from *iWoz—How I Invented the Personal Computer, Co-founded Apple, and Had Fun Doing It.*

I'm not sure how or when the interest began, but it started when I was very young. It may have something to do with the fact that I was born during the International Geophysical Year, within a year of Sputnik's launch. Or it may have been the sunspot cycle—I was born at the peak of cycle number 19, one of the strongest peaks on record.

I must have been around seven years-old when I managed to get a little electrical timer working for my grandmother. She was still living in the Bronx, and she worried about burglars. When I got the timer to actually switch the living room light on and off, she declared that I was "electrically inclined." I guess she was right.

It may have been the Boy Scout handbook that first alerted me to the joys of electrical experimentation. Perhaps prompted by this book, on trips to the Congers, N.Y. hardware store I would ask mom to buy me the batteries, small bulbs, wire and switches to construct a simple electrical lighting system.

Several key birthday and Christmas presents had an influence: There was kind of an electronic version of an erector set—it had plans for a short-wave radio, but the assembled parts never seemed to pull any signals in. There was a tape recorder—I really enjoyed recording voices and playing them back. And one year during our beach vacation at Lavallette, N.J. someone brought along a little wireless intercom system that allowed us to speak among the many bungalows that our large group of family and friends had rented—for some reason I really liked that little intercom…

Another influence had to have been Apollo. I was ten years old when they landed on the Moon, and I was completely captivated by the space program. We all had GI-Joe toys, but I had the astronaut version (complete with a very cool Mercury-style capsule). I was so interested in the space flights that I would convince my parents to let me stay home from school on days in which particularly interesting things would be happening in the moon shots. (I realize now, of course, that skipping school—no matter how scientifically noble the motivation—was probably not a good move for an aspiring astronaut.) I converted a closet into my own flight simulator and at one point pledged to spend days cooped up in there—I wanted to test my ability to withstand the confinement of the Command Module.

My interest in electricity yielded a science fair project: With the wires and buzzers and batteries that I'd acquired from the hardware store, I built a "Question and Answer Machine." A TV-sized box had a front panel with written multiple choice questions (I think they were about science). Each of the possible answers had metal bolts beside them. The device had an "answer probe" that you touched to your answer. Get it wrong and you got a nasty loud noise from the buzzer. The electrical design was all mine, but dad helped me build the cabinet. The whole thing looked so good that I think the judges thought that it wasn't really my work. I suppose there were what could be considered some rudimentary logic gates in that thing. It's a shame that the school didn't give me any encouragement in this area.

But I did get a lot of encouragement from my parents. I remember my mom saving the plastic wrappers from the rolls of paper towels so that we could get an encyclopedia. And she bought me a book that had a real impact on me: *The Amateur Scientist* by C.L. Strong. This was a compilation of articles from the Amateur Scientist column of the magazine Scientific American. "Build a Homemade Atom Smasher!" "How to Tranquilize a Rat." "A Homemade X-Ray Machine." Any one of these articles would today provide a veritable gold mine for litigators. Luckily, I didn't do any atom smashing or X-raying, but I did build the non-lethal rodent trap described in the book. (Decades later, on an island in the North Atlantic, my children and I would use this design to capture Azorean lizards.)

There were a number of what I guess can be called precursor hobbies. Apollo generated an interest in model rockets. I practically memorized the Estes company's catalog. My favorite rockets were the Astron X-Ray (it had a transparent payload compartment that was perfect for testing the effects of high G forces on small animals) and the Astron Big Bertha (it was big, and it seemed to rise more slowly from the launch pad, reminding me of the Saturn V's majestic ascent). Trying to emulate Von Braun, two guys from the neighborhood and I formed the "Waters Edge Rocket Research Society," and almost instantly discovered that in so doing we had crossed the nerd line. The less astronautically inclined kids (who were all destined to be on the football team) taunted us mercilessly.

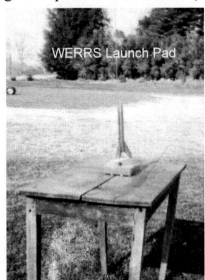

Of course, there was also astronomy. I looked at Jupiter and Saturn and the moon with a variety of small telescopes.

We can't forget amateur photography. As would later happen with model airplanes, finances put limits on technical progress. I could afford just about all that was needed for my darkroom… except an enlarger. So in the little photo lab that I installed in our downstairs bathroom, I proudly developed black and white 35-millimeter prints, prints that were exactly 35 millimeters wide.

There were model airplanes. I built a balsa wood plane kit that was rather cruelly called the FUBAR 36. (I didn't know the meaning of the acronym.) Radio control equipment existed, but I couldn't afford it, so the FUBAR was flown in what was known as "free-flight" mode. There was this pitiful little burning fuse arrangement that was supposed to pop the rear stabilizer into a vertical position and bring the plane to a soft landing, but this was all very theoretical. Flights involved lots of frantic chasing. The plane lived up to its name.

And for a while I was crazy about real airplanes. I'd ride my bike up to the DeForest Pharmacy and spend far more than the acceptable browsing time looking at "Flying" magazine. Longer bike trips would get me to Ramapo Valley Airport, where I'd longingly watch the Cessnas and the Piper Cubs come and go. I must have been around twelve when I decided to save up for a plane ride. My dad was nervous about this and wouldn't let me do it. But, as always, he supported my interests. He learned that there was a World War I replica aerodrome way upstate in Rhinebeck, New York. He and my mom gallantly hauled all five kids up to Rhinebeck so that I could see the show. Eventually he got more comfortable with the idea of a flight for me—on my thirteenth birthday my dad bought me my first ride in a plane.

I didn't know it at the time, but looking back, I clearly had the not-so dread disease described in the epigram at the beginning of this chapter. I know now that I had Dilbert's disease. I had The Knack. All these other interests were leading me to amateur radio.

UNDERSTANDING: ELECTRONS AND ELECTRICITY

"All things by immortal power;

Near or far,

Hiddenly

To each other linked are,

That thou canst not stir a flower

Without troubling of a star..."

From "Mistress of Vision" by English poet Francis Thompson (1857-1907) as quoted in "Why Antennas Radiate" by Stuart G. Downs, Jan/Feb 2005 issue of *QEX*

Like most kids who play with batteries and wire, I wondered what it was that made those buzzers buzz and light bulbs light. Consultations with adults and library books provided some superficial explanations. I was told that it was electricity that did the job. Perhaps I even learned that small objects called electrons were pulled from atoms and made to run through the wires. But I was left wanting to know more. Later, when I was studying to become a radio amateur, I felt an *obligation* to know more, to really *understand* what an electron was, and how it managed to run through the wires.

I didn't realize it at the time, but in seeking deep levels of understanding, I was setting some unreasonably ambitious goals, and setting myself up for a collision with some of the fundamental mysteries of the universe. Richard Feynman won the Nobel prize for his development of the theory of Quantum Electrodynamics—the scientific explanation of how electrons interact. Here's what he says about trying to *really* understand these things:

> "I think it is safe to say that no one understands quantum mechanics. Do not keep saying to yourself, if you can avoid it, 'But how can it be like that?' because you will go 'down the drain' into a blind alley from which no one has escaped. Nobody knows how it can be like that."

Isaac Asimov put it this way:

> "Exactly what an electric charge is we cannot say. However we can say how a substance with electric charge acts, and how we may measure the extent of this action and therefore the size of the electric charge. This is an operational definition of electric charge and it is enough to satisfy scientists at least for the time being."

In 1989 a British engineer named F. A. Wilson published a wonderful book entitled *From Atoms to Amperes*. It is filled with the kind of wisdom and advice that I could have used during my days as a teenage radio fiend. The book was published by Bernard Babani Ltd. of London and is unfortunately out of print. (It is so good that I think we should launch a campaign aimed at getting the publisher to bring it back.) My copy is now held together (appropriately) by duct tape, crazy glue, and rubber bands. The yellowed pages are scarred with my underlines, circled phrases, and exclamation points, all marking the many Eureka moments that F.A. Wilson provided.

The opening chapter of the book is entitled "Avoiding the Deep End," and begins with this very useful caution:

> "We are not physicists whose mission in life is to dig deeply into the mysteries of the universe but people wishing to know something about what makes electricity tick. It is as well therefore to decide first just how deeply we wish to get involved."

That's a bit of wisdom that I wish had been included in some of the books that I was studying as a kid. Those books often were quite smug, and left me with the impression that it was possible to completely understand—at a deep level—the nature of the electron. I felt that somehow I was just missing something. It wasn't until years later, when I came across Feynman's honest and humble admissions, and Wilson's good advice, that I was able to accept the fact that there are limits to our understanding (and certainly to *my* understanding) in these areas.

So, in an effort to stay out of Feynman's blind alley and Wilson's deep end, I decided not to torture myself trying to understand what the electron is made of, or where its electrical charge comes from. I essentially went back to the very simple model of the electron that they teach you in school—little bits of charged matter swirling (somehow) around the nucleus. As Wilson put it, "... the work of the battery is to separate the electrons from their atoms and direct the differently charged particles to the appropriate terminals..." As for the wire, "When a metal allows electrons to flow through it easily, it is said to have high conductivity. This is due to its having a copious supply of free electrons available, free in the sense that they are not permanently attached to atoms, but are free to move around."

After a while, I was no longer beating myself up about my incomplete understanding of the electron. I learned to live with it. I told myself that the true wizards of the radio art would not look down on me for this gap in my understanding. After all, I reasoned, carpenters aren't expected to understand the molecular interaction of nail and wood. Cooks aren't expected to understand the chemistry underlying their sauces and soufflés. And so I shouldn't have to *really* understand the electrons that move through our circuits. But still, it kind of bothered me…

****

In the suburban town that we had moved to in Rockland County New York, there was an odd little building on Route 303 that somehow attracted our attention. It was off all by itself, with the highway to the front and woods to the back. It had a sign above the door: "W2DMC – CRYSTAL RADIO

CLUB." My four siblings and I made that intriguing little building part of our family travel routine. As we'd ride around in the back of our Plymouth station wagon with mom at the wheel, every time we passed in front of that place, one of us would call out "There's the radio shack!" I always wondered what was inside.

My father was a New York City policeman. He spent his 26 years on the force in New York's 34th Precinct. "The Three Four" in cop talk. As far as I was concerned, "New York City Cop" was about the coolest thing a dad could be. I actually felt sorry for kids whose dads were bankers and lawyers and dentists… My dad carried a gun and went out each day to protect people from bad guys. My dad's friends were extremely cool and interesting. Once, my brother and I were delighted when we woke up and found an enormous motorcycle in our garage—my father's hung-over cop friend was asleep on our couch, having wisely decided not to tempt fate on the roads the night before. We occasionally got to go with my dad to the cop bars, where we sat (often up *on* the bar) and listened to amazing tales of cop derring-do.

As a cop, my father worked "around the clock," following a mysterious schedule that only my mother seemed to understand. For a few days he'd work the normal hours of 8 a.m. to 4 p.m. Then he'd switch to 4 p.m. to midnight. Then would come the "late tours" of midnight to eight. I haven't actually crunched the numbers, but I think this kind of schedule results in more dad-kid interaction than your standard 9-5 life. My dad would very often be home when we got back from school. (Somehow he'd developed the ability to sleep through the mayhem that five young kids can create in a wood-frame "raised ranch" house.)

Dad listened to AM radio, and he liked to cook. I'd often find him sitting in our kitchen, slicing up carrots or something, with "WOR, 710 on the AM dial" tuned in. I think it was this station that turned my vague "electrical inclination" into something far more serious.

My father's favorite radio personality was Jean Shepherd. "Shep" as he was known to his fans, came on each night. His theme song was "The *Bahn Frei* Polka" by Strauss. His slogan was "EXCELSIOR!" Shepherd was a very talented storyteller. He published wonderfully funny books with titles like *Wanda Hickey's Night of Golden Memories*, and *In God We Trust—All Others Pay Cash*. Each night, as my father chopped the carrots, Shepherd would reach into his seemingly bottomless sack of stories. He had to do a show each night, so I suppose he often had to dig deep.

From time to time, Shepherd would turn to technical topics. As a kid he'd become obsessed with amateur radio. With the depression in full swing, there wasn't a lot of money lying around to indulge the technical whims of adolescent boys, so Shepherd and his electronic co-conspirators had to build their own gear. Over WOR, Shep described the adventures that he and Stan and Bollis had in procuring needed parts and then trying to build their own radio gear. And he conveyed the spirit of brotherhood that develops amidst the solder smoke and musty boxes of old parts. Shep's stories made me want to join the brotherhood.

The "Electronics 59" store in nearby Spring Valley was a place where I could put my hands on real radio stuff. And they had real ham radio magazines. I picked up my first copy of *73* there. I was probably already hooked, but that magazine set the hook in a lot deeper. The writing was heavily laced with the slang and acronyms of ham radio (a style I'm trying to avoid in this book), so I couldn't understand half of it. But the magazine presented an updated version of the 1930's world of intrepid, technically skilled enthusiasts described by Shepherd.

It became clear that the first thing to do was simply listen in on the fun. I needed a receiver. I started poring over the back pages of the magazines and the catalogs, and after an agonizingly exhaustive search, concluded that the receiver for me was the Lafayette HA-600A General Coverage short-wave receiver.

The Lafayette had many virtues. "General coverage" meant that it could cover all of the short-wave frequencies, not just those portions of the spectrum used by radio amateurs. This meant that I could use it for short-wave listening—I could tune into the BBC and Radio Moscow and other short-wave services! I could listen to ships at sea and aircraft crossing the Atlantic. There was no doubt about it. I desperately needed a Lafayette HA-600A. I was probably 14 years old at this point, so I was way beyond letters to Santa. Instead I just showed my mother the by now dog-eared page from the catalog. I don't know how she managed to find one of these things, but, of course, she did—the coveted Lafayette was under the tree that Christmas.

That little box of Japanese transistors opened up a new world for me. I spent most of that Christmas vacation bounding up the stairs to let mom and dad know that I'd picked up signals from yet another amazingly distant and exotic location. I doted over that radio. I lived in fear that some sort of freak mid-winter lightning bolt would reach down from New York's leaden skies and destroy my receiver. I worried that dust would somehow clog its vaunted "jeweled movements" (whatever they were), so I covered the receiver with a little cloth when it was not in use. My younger brother Ed (who shared the room with me and who was by now doomed to years of trying sleep through my late night radio adventures) snickered at all this.

The short-wave listening was great. I especially like the programs of HCJB, the Voice of the Andes in Quito, Ecuador. That station actually had programs about short-wave listening and ham radio. In a world of 12-channel TV and no internet, programs that were focused on my very esoteric interests seemed miraculous indeed. But as I listened to that HA-600A, it was the amateur radio chatter that really caught my attention.

The short-wave frequencies are generally considered to be those between 3 and 30 megahertz. Hertz means cycles per second. Mega, of course, means millions. A signal at 3 megahertz (3 MHz) is a vibration of electromagnetic energy (more on this later) that is going through 3 million vibrations each second. If you factor in the speed of light, it turns out that as these signals move out from the antenna, there is about 80 meters (metrically challenged readers can think of 80 yards) between each vibration.

By modern standards, these waves are not at all short. They are actually very long in comparison to the waves that are emitted by your computer's Wi-Fi router, or your cell phone. But radio was very new when the term was coined, and at the time it was believed that only very long waves (more than 200 meters) were useful for long-distance communications. Amateurs were relegated to the apparently useless short-wave frequencies. Only later was it discovered that on these frequencies something really marvelous happens.

Radio waves normally travel in a straight line—"line-of-sight." This represents a real problem if you are trying to communicate with someone on the other side of our round planet. Even if you aim at the horizon, "line of sight" means that your radio signals are going to be flying out into space. Reaching that fellow in Australia is going to be a problem. But then they discovered a mirror in the sky.

As they experimented with their homemade gear on the frequencies that no one else wanted, radio amateurs in the early 1920s noticed that they were making contacts with stations that were far beyond line-of-sight range. In fact, they soon found their signals crossing the Atlantic, and (a bit later), the Pacific. It took a while for someone to come up with an explanation: British scientist Oliver Heaviside theorized that an ionized layer of the earth's atmosphere reflects (actually it refracts) radio signals in a certain range of frequencies. Fortunately for radio amateurs, this reflection takes place in the short-wave bands to which we have been sent off to play.

So, some fifty years after this happy discovery, I sat in my bedroom late into the night with headphones attached to my beloved Lafayette HA-600A, listening to signals bouncing in from around the globe.

The short-wave broadcast signals (BBC, Radio Netherlands, etc.) were wonderful, but it was the ham radio chatter that really captivated me. Once the long-range communication potential of short-wave had been discovered, commercial and government operations had moved into the amateurs' playground, but fortunately we had held on to some small but useful bits of the short-wave frequency spectrum. Our electromagnetic territory was now a set of frequency bands: We could operate, for example, from 3.5 million cycles per second (MHz), up to 4 MHz. Referring to the physical length of the waves, this was the 80 meter band of frequencies, the upper portion of which is sometimes called 75 meters.

Within that 75 meter band, at around the frequency of 3.885 MHz, I came across the brotherhood of ham radio. The voices were young—these guys were not much older than I was. Most were obviously in college, but some seemed to still be in high school. They seemed to be scattered across New England and the Mid-Atlantic states. In more ways than one, it was technology that brought them together; it was the discoveries of Faraday and Maxwell and Hertz and Marconi and Fessenden that allowed them to meet and communicate, communicate not in some college bar or on some gum-speckled suburban corner outside a Burger King, but "ON THE AIR." In a deeper sense, technology was the foundation of their fellowship—it was their common love of electronics that brought them together. Listening to them, you could almost smell the solder smoke. This seemed to be Shepherd's old gang, reborn in the 1970s. Technical ability defined them as a group, and provided a kind of hierarchy within the brotherhood. The radio gear allowed them to talk, and when they talked, they talked about radios—about building radios, and fixing radios, and modifying radios. The medium was the message.

Of course, many of us had gathered around this technological campfire because we'd been shunned by other groups. For us, amateur radio was a refuge from the social isolation felt by kids who were good with soldering irons, but not with footballs. Decades before the internet, ham radio offered the same kind of escape that the net would eventually provide for those kids who had not made it into the ranks of the "in-crowd." The cruel social world of high school offered rejection and humiliation, but with ham radio, acceptance and friendship were available 24/7 via electromagnetic waves.

Every group of young males seems to require or generate an enemy, an "other," the outsiders, the non-us. I suppose the football player/cheerleader crowd could have filled this role, but I guess a technology-based group needed a more technologically defined enemy. And in the 1970s, one was available, good buddy.

Yes, the enemy was the CBers, the Citizens Band operators. Listening to the 75 meter gang and reading the ham radio magazines, I learned that outside the hallowed ranks of ham radio, there was a vast wasteland of unwashed radio barbarians. For these people, radio was a toy, a fad. They plunked down their fifty bucks and, after struggling to figure out where to plug in their store-bought antennas, they flipped through the CB channels and started yakking away in their pathetic "10-4 good-buddy" radio patois.

Technical ability was what distinguished us from this rabble, and our technical skills had government certification. We had licenses, and we had to pass very difficult exams to get them.

It was the difficulty of the exam that made the amateur radio license attractive. Here was something that was hard to get even for adults, but available to you even if you were a 13 year-old kid. If you passed the exam, you became a government-certified expert, and a member of the fraternity. You were issued a unique call sign that you could tack onto your name. (Jean Shepherd said that at age 14 he thought of himself as "W9QWN, A Man of Substance.") I eventually became Bill Meara, WB2QHL. For me, that call sign carried the same weight as Ph.D. They even listed you in this big, phonebook-like directory of U.S. Radio Amateurs. Getting an amateur radio license was a very big deal for a 13 year-old kid in early 1970s.

The Lafayette HA-600A (with jeweled movements) continued to provide motivation. I soon became a regular listener to the conversations of that group of young enthusiasts on the 75 meter band. I was captivated not just by the good fellowship, but also by the fact that their signals were among the easiest to tune in. These guys were transmitting on AM—Amplitude Modulation. Their "rigs" were relatively simple compared to the Single Sideband (SSB) transmitters that had come into vogue during the 1960s. While my beloved receiver could tune in the SSB signals, if you didn't tune them in just right they would have a Donald Duck sound to them. But the AM was always very clear and strong. There would always be a big "ker-chunk" sound when one of the boys switched his station to the transmit mode. The carrier signal would immediately overwhelm the background static, and for a second or two before he spoke, the ether would be quiet. They sometimes seemed to linger a bit before speaking, perhaps to let listeners marvel at their domination of the airwaves.

Most of these guys were using big, heavy commercially manufactured (but often kit-assembled) transmitters like the Heathkit DX-100. These rigs weighed in at around 100 pounds. They were filled with hot vacuum tubes and deadly high voltage. There was usually 800 to 1000 volts DC inside those cabinets. The DX-100 had a kind of wire mesh top to it—you could peer down at the final amplifier tubes and look right at the plate voltage caps on top of those 6146s, knowing that there was enough juice there to instantly add your name to the rolls of what the ham radio magazine *QST* called "Silent Keys." Of course, for teenage boys, the danger was part of the attraction. We were playing with serious machines.

By now I was hooked, but I needed help in joining the fraternity. The ham radio books and magazines that I had accumulated advised getting in touch with a ham radio club. That little radio shack on Route 303 seemed to be the place for me to go. Somehow I found their meeting schedule, but I was 13 years old and shy. My dad volunteered to take me up to one of the regular Tuesday night meetings at the clubhouse. But when we got there, I got cold feet and suggested that we forget all about ham radio and go back home. But dad walked in with me, explained my interest to the club members, and asked them to help me out. I was in! Dad returned to our station wagon and patiently read his Daily News while I sat through my first meeting at the Crystal Radio Club. Finally I was inside that mysterious radio shack that we'd been driving by for years.

Amateur radio clubs around the world seem to share some characteristics: They always seem to have piles of old ham radio magazines, and a mild, not-unpleasant smell that comes from having these magazines exposed to humidity for many years. There are usually some piles of electronic junk and assorted parts. Bulletin boards with long-outdated announcements of flea markets and summer picnics. Uncomfortable folding chairs. Tobacco residue. Inadequate ventilation, heating and/or air conditioning. I suppose it is a good thing that these places aren't too comfortable—if they were, members might start spending marriage-threatening amounts of time there amidst the beloved parts and magazines.

I think the demographics of radio clubs are also the same around the world. In spite of efforts by some of the national-level radio organizations to pretend otherwise (they seem motivated by a combination of PC concerns and recruiting hopes), this is definitely a guy thing. The group I met at the Crystal Radio Club had a wide variety of professions—we had dentists and lawyers and plumbers—but there were no women in the group. Most of the members were middle aged (old in my eyes) but there were a few people my age. In a reminder that we were still dealing with a very young technology, there was one old timer who was a charter member of the Crystal Radio Club; when he'd joined, crystals—small chunks of galena rock—were what cutting edge hams used to listen to radio waves.

But as the magazines had promised, I quickly found that an interest in radio was all that was needed to break into this kind of group. The Crystal Radio Club welcomed me to their ranks.

The club was a lot of fun. I looked forward to the meetings, even the rather stuffy and formal monthly business meeting (complete with a reading of minutes and the use of Robert's Rules of Order). We all also met once a week "on the air." Each June we participated in a nation-wide emergency preparation test called "Field Day." We took to the field with tents and generators and set up multiple radio stations at an abandoned Nike Hercules missile site. We had club outings: We went up to the hallowed halls of the Newington, Connecticut headquarters of the American Radio Relay League (ARRL), and got to visit the famed station W1AW (the station that had helped me learn Morse code.) We went to New York City for the 1974 national convention of the ARRL.

There is a very nice tradition in amateur radio of having a more experienced amateur agree to help out a newcomer. Somewhere along the line, the mentor became known in ham radio circles as the "Elmer." (There had probably been some guy named Elmer who been famously helpful to new hams. Three cheers for Elmer!) Mine turned out to have a name that sounded remarkably like Elmer—my Elmer was Hilmar.

Hilmar Maier was an immigrant from Germany. He lived not far from us (within bicycle range) and had kids who were a bit younger than me. Hilmar was himself relatively new to ham radio, but he was a professional electronics technician, with a very high level of technical skill and knowledge.

Having linked up with a club and having my own Elmer, I began to study for my first ham radio license. At the time, the Novice class license was kind of an electromagnetic learners' permit. You had to pass a relatively simple exam on technical matters and radio regulations, and you had to demonstrate the ability to send and receive Morse code at five words per minute. If you passed the test, you would be allowed to venture out and transmit on small, restricted portions of the amateur radio frequency bands, using limited power, and Morse code only. And the license would be good for just one year—you had to upgrade to a higher class of license, or go off the air. With a novice license, you'd be a ham, but just barely, and with an expiration date. You'd get a call sign from the FCC, but it would have a big N letter in the prefix. I became WN2QHL. The N let the radio world know I was a novice.

UNDERSTANDING: RADIO WAVES

"You see, wire telegraph is a kind of a very, very long cat. You pull his tail in New York and his head is meowing in Los Angeles. Do you understand this? And radio operates exactly the same way: you send signals here, they receive them there. The only difference is that there is no cat." —Einstein

My studies for the novice exam had given me some exposure to the basic physics of electromagnetic waves. Once again, without realizing it, I was colliding with some of the deepest mysteries in physics, and once again these mysteries collided with my teenage desires to become a true radio wizard. I really wanted to understand what radio waves ARE and how they are generated. After all, this is the core technology of the radio amateur. How could I look myself in the eye and call myself a true ham if I didn't REALLY understand what radio waves were?

My 1973 edition of the ARRL Handbook seemed to kind of breeze past the physics of radio waves, devoting less than one page to the subject before quickly moving on to the practical topics of antenna construction and ionospheric propagation. Some of the explanations that were provided at first seemed to make sense, but then crumbled after just a little thought. For example, one book explained that radio waves propagate through space because they do not have time to collapse back into the antenna before being pushed out into space by the waves behind them. This sounded good, but, uh, what about the last wave? What's pushing that one out?

Once again, I was left with the impression that I was just missing something that others had come to understand. Books often implied that only those with highly developed mathematical skills could hope to understand radio waves; it was only much later that I came to realize that there are limits to understanding on this subject, even for the mathematically skilled. Writing about antennas in his excellent and very mathematical book, *The Science of Radio*, Professor Paul Nahin of the University of New Hampshire notes, "The concept of energy flowing through free space is really quite an astonishing claim. We can write equations all day long showing mathematically how it goes, but if you try to form a physical image of such a flow, you will fail (at least I fail!)."

Paul Nahin's candor and the admonitions of Richard Feynman (about blind alleys) and F.A. Wilson (about the deep end) all would have helped me in my teenage quest for understanding—it would have been comforting to know that there are limits for everyone in this area. But I suppose it was a good thing that I did not know about these limits, because I just kept on digging, and in the process came across some inspiring people.

The English inventor Michael Faraday played a central role in the 19th century discovery of radio waves. I think Faraday should be a hero for radio amateurs, for in many ways he was one of us. Faraday was from a poor family, and he was self-educated. He worked in a book-binders shop, and in his spare time read the books that he was assembling. In her excellent book *Einstein's Heroes—Imagining the World Through the Language of Mathematics*, Robyn Arianrhod wrote:

> "Faraday soon became captivated by the Encyclopedia Britannica's detailed entry on electricity, which fired his imagination to the extent that every day he hovered longingly in front of an old rag shop, until he could afford the seven pennies needed to

buy two old jars that, with the aid of a bullet and a bit of wire, he turned into his first electrical apparatus."

He was unschooled in mathematics. He was very much a hands-on, in-the-workshop kind of guy. Yet he made the discoveries that led humanity to radio waves. Faraday discovered the links between magnetism and electricity. He came up with the concept of electric and magnetic fields.

Faraday's key discovery was that a CHANGING magnetic field creates electric current in a nearby wire. The word *changing* is the key. It really wouldn't make sense for a static magnetic field to generate electric current. If this were the case we'd simply need to put wires near magnets and then turn on the lights. But that, of course, gets us into the area of perpetual motion machines and violates the laws of physics. Faraday discovered that for energy to be transferred, those magnetic fields need to change. All of our power plants are built, in essence, to spin magnets around wires (or vice versa)—to create the changing magnetic fields that generate electric current.

Faraday made the key discoveries, but he didn't have the skills to put what he was seeing into mathematical form. Enter James Clerk Maxwell. For different reasons and in other ways, Maxwell too should be an inspirational hero for radio amateurs.

Maxwell's mother recalled that as a young child, he was happily curious about the world. He would always be asking his parents about how things worked: "What's the go of it?" and (following up) "What's the *particular* go of it?" or "Show me how it *doos*!" In what was perhaps a premonition of things to come, as a child Maxwell was fascinated by the bell-wire signaling system in his house (perhaps used to summon servants). Unlike Faraday, Maxwell did not come from a poor family—he'd had the benefit of education at the top universities of Scotland and England. But—like many of us—he was clearly something of a nerd, and suffered at the hands of bullies. Arianrhod described him this way:

> "Shy and slow of speech, he wore strange clothes—designed by his father with comfort rather than fashion in mind—which made him such a target for the other students that he arrived home from his first day at school with his clothes in tatters."

Of course, as many of us know from personal experience, Maxwell wasn't the only technically inclined kid to suffer in a school yard. Witness Marconi's experience:

> "Electricity became a fascination for Marconi early in his childhood… Marconi was possessive about electricity. He called it 'my electricity.' His experiments became more and more involved, and consumed increasing amounts of time." (From *Thunderstruck* by Erik Larson.)

> "Young Marconi grew up as something of an outcast, speaking Italian badly and with an exotic and obviously well-to-do foreign mother. This did not endear him to his contemporaries at school and after a miserable period he was taken home to study under a private tutor." (From *In Marconi's Footsteps—Early Radio* by Peter R. Jensen.)

Maxwell was young when Faraday was old, but the Scot had the tact and people skills needed to bridge the generation gap and to build a good relationship with the old man whose discoveries he wanted to explore.

Maxwell came up with the equations that describe the interactions between magnetic and electrical fields, the interactions that Faraday had first discovered in his London workshop. But Maxwell took it further, and, in an excellent illustration of the power of mathematics, saw in his equations something that no one else had seen: In those equations, in his mind's eye, Maxwell saw radio waves.

It is important to understand that radio waves were discovered not in a workshop or a lab, but in mathematical equations contemplated in the mind of James Clerk Maxwell. In a very real sense, the waves popped out of the equations. Only later did Heinrich Hertz build an apparatus that generated the waves called for by Maxwell's formulae. (Marconi got the idea for using radio for communications purposes when, at age 20, he read Heinrich Hertz's obituary.)

Maxwell's equations tell us that a changing magnetic field creates a changing electrical field. It will create this changing electrical field not only in a nearby wire but—and this is important—it will also create a changing electrical field in free space, in a vacuum. Another of his equations says that this changing electrical field will create a changing magnetic field, also in free space. You can see where we're going with this: As far as you can go, and at the speed of light.

In Maxwell's day it was thought that just as air carries sound waves, there had to be some sort of medium carrying the electromagnetic waves. They called this medium "the luminferous ether." But then two Americans named Michelson and Morley decided to do some tests to see if they could detect the ether. They reasoned that the earth's motion through space should affect the speed of light arriving to us from different directions: If the earth is moving rapidly in its orbit toward an object, light moving through a fixed ether should be coming at us faster than light coming from objects in the opposite direction. They used a large interferometer for their tests. They discovered no difference in the velocity of light. The speed of light was not affected by the movement "through the ether." The speed of light is always the same. That's very weird.

Although Michelson and Morley proved that ether doesn't exist, we know that free space has measurable electrical characteristics. It has specific values of Permittivity and Permeability. These values determine how much of an electrical field will be created in free space by a changing magnetic field, and vice versa. The amount of change experienced by a given region of free space depends on how quickly the electrical and magnetic fields are moving through it. Maxwell, working with his equations, discovered that there was only one speed for these waves that would not violate the law of conservation of energy: That speed was c, the speed of light.

By observing the orbit of Jupiter's moon Io, the Danish scientist Ole Romer had calculated that light from Jupiter takes about 18 minutes to get to earth. He did this in 1676. But Romer did not know the size of Earth's orbit, so he was unable to actually calculate the speed of light. In 1849 Hippolyte Fizeau used an ingenious system of cog wheels and mirrors to calculate that speed. Reflected sun light was sent through a cog wheel to a second mirror several thousand meters away. Fizeau increased the speed of the cog wheel until—at a certain speed—light that went out through one gap could be seen coming back in the second gap. Knowing the distance traveled and the speed at which the wheel was spinning, Fizeau was able to calculate c. Bravo Fizeau!

When Maxwell saw the same speed figure popping out of his equations for electromagnetic waves, he became the first human being to know what light is. Light is an electromagnetic wave at a frequency for which our eyes have evolved sensitivity. Our eyes are natural, biological, electromagnetic wave receivers, created over millions of years by evolution. I guess it should come as no surprise that we are interested in building radio receivers, because in a certain sense we ARE radio receivers.

****

For me, the Morse code restriction was the most significant downside of the novice license. I'd been attracted to ham radio by listening to hams TALK on the radio, talk with microphones, not with telegraph keys. I wanted to join in those conversations in which you heard the other guy laugh and tell jokes, in which you recognized the voice of a friend calling in from across the country or from the other side of the world. It was hard to imagine doing this through a series of dots and dashes. But that was the requirement, so I began the process of learning Samuel Morse's code.

With some guidance from the club members I basically pounded the code into my head. I had asked for a tape recorder for my birthday (further proof of electrical inclination), and found that it was very helpful in the campaign to learn Morse code. The trick is to get your brain to automatically associate certain sounds with corresponding letters. You had to make it audible. You had to get your brain to associate the sounds of dots and dashes with the letters. It wouldn't work if you just tried to memorize the chart showing the alphabet and the corresponding dots and dashes. You couldn't even think of them as dots and dashes—you had to think about them in terms of their sounds: dits and dahs. This was more training than memorization. Using a little buzzer, I'd practice sending the letters while recording them. Then I'd play the tape back and work on my ability to receive them. (My mother had to listen to a lot of this, and to this day can recognize a few letters in Morse.) The big breakthrough came when I could finally understand what ham radio operators were saying to each other in their on-the-air Morse code conversations. Then it became fun, and a real source of pride. Many adults had serious problems learning "the code," but hey, hey, hey, I was now getting good at it.

Weeks and weeks of practice with the tape recorder and the HA-600A (with jeweled movements) got me closer and closer to the five words per minute needed to pass the Novice exam. I was also helped by the Morse code practice broadcasts of radio station W1AW from Newington, Connecticut. W1AW was the official headquarters station of the American Radio Relay League (ARRL), the national association of radio amateurs. For a 13 year-old kid, on the importance scale, W1AW and the ARRL seemed to be right up there with the Pentagon, the White House, and CBS News. Strictly following a schedule published in *QST* magazine, W1AW's powerful signal would appear at the appointed spot on the radio dial, and would solemnly begin to transmit flawless Morse code at scientifically determined speeds. On cold winter's nights I'd be sitting there in Congers, headphones on, hunched over my beloved HA-600A (w.j.m.), pencil and paper at the ready. In an effort to improve the speed and clarity with which I wrote the incoming letters down, I actually modified my handwriting, abandoning the beautiful cursive script that the nuns at Good Shepherd and St. Paul's Elementary Schools had drilled into me; I adopted a print style that seemed to get the letters onto the page faster. I've never gone back to cursive. Morse code ruined my handwriting. I hope the nuns are not reading this.

While learning the code I studied for the technical exam. The fact that you had to pass an exam seemed to highlight the power of radio. You were seeking access to a technology with global reach! And the authorities of the FEDERAL GOVERNMENT did not want you to be inadvertently screwing up the global communications system: One false move from you and airliners in India might go off course, SOS calls from the South Pacific might not be heard… Shortwave radio was serious business. Only the technically competent would be allowed to participate.

The guys at the club helped me prepare for the test. There were some comic moments. I seemed to come up with some pretty unique pronunciations of the technical terms that I read in the ARRL Handbook. I must have really mangled "quiescent" (as in a transistor that is, for the moment, not amplifying) because that really had them scratching their heads. But the real eye-popper came when I asked them to explain the concept of impedance (the total resistance to alternating current flow in a circuit). The correct pronunciation (I learned) was imp-eeeee-dance. Unfortunately I was asking these middle-aged guys about what sounded to them like im-PUH-dins (an entirely different technical problem). I bet that was the first (and last) time THAT subject came up at the Crystal Radio Club of Valley Cottage, N.Y. Needless to say, my pronunciation was quickly corrected, and the electronic concept was very kindly explained.

The novice "ticket" had a definite learner's permit feel to it. Even the setting for the administration of the novice exam made it feel like a radio license with training wheels: instead of having to go to a federal government building, the novice exam was given by a ham radio volunteer, in my case it was given by my Elmer, Hilmar.

I passed the test, and after what seemed like an eternity, on April 27, 1973 my license arrived in the mail. I was now WN2QHL. Now it was time to finally get on the air.

I already had the needed receiver—now I needed a transmitter. Through the club, I bought a Heathkit DX-40 transmitter, a low-power version of the famed DX-100 that many of the guys on 75 meters were using.

Heathkits were wonderful. They seemed to be designed to go to market at a reasonable price. If you bought a Heathkit new, you got a big, carefully organized box of parts, and very, very detailed instructions on how to assemble your new device. (Try to imagine people today buying their computers this way.) Weeks or months of painstaking construction (followed perhaps by weeks of frustrating troubleshooting) were needed to get the new rig going. Thank God my Heathkit was second hand—at age 13 I think I might have died of impatience if I'd actually had to build my own.

The DX-40 was a very simple transmitter. It used vacuum tubes, not transistors. It transmitted Morse code, but was also capable of transmitting Amplitude Modulated (AM) voice. Novices, of course, were limited to Morse, but the fact that my transmitter actually had a microphone jack on it served as an enticement for license upgrading. The DX-40 was "crystal-controlled"—the transmitted frequency was determined by a plug-in piece of quartz that had been cut for one particular frequency. While my receiver could tune up and down the dial, on transmit the DX-40 was (in a very apt phrase) "rock bound." Crystal-controlled transmitters had been common among hams in the early days, but by the time I got involved most amateurs had acquired the technology needed to vary their transmit frequency. In fact, by this time most hams were using "transceivers," i.e. combination transmitter-receivers in which the transmit frequency was automatically the same as the frequency being received. Crystal control may have been practical decades earlier, but in 1973, it was seriously outmoded. If you heard a station that you wanted to talk to, unless you were phenomenally lucky, his frequency would be different from that of your crystal. (Crystals were expensive, so you'd have only a few.) You could try to call him on your crystal frequency, but odds were that he would not be tuning around to try to find you—accustomed to talking to other guys with transceivers, he would only listen on his transmit frequency. That's why "rock bound" was such an apt phrase—you'd be like Robinson Crusoe trapped on his island, waving frantically at passing ships that weren't bothering to look for him. I eventually got off that little island by acquiring a Variable Frequency Oscillator (VFO)—in my case a very cool-looking Globe Electronics "VFO Deluxe."

I also needed an antenna. I needed something that would allow the radio frequency energy from the DX-40 to zip up and down along a wire and emit electromagnetic waves. How this happened was a mystery to me; and to some extent it still is, but at the time I was focused on getting on the air, so I put the deep physics questions aside, and with my nose in the ARRL Handbook decided to build a dipole antenna for 40 meters.

The radio books depict dipoles as simple, happy antennas. Just two lengths of wire each one quarter wavelength long, suspended at each end by a friendly, convenient tree and with a bit of coaxial cable attached at the center, in the radio books the dipole always seemed to be simplicity itself, two little bits of wire in an inverted arc, smiling out at the universe from high above the radio shack.

The reality of course, was very different. You need a good technique for attaching the coaxial cable securely to the antenna. And you needed a method of getting the two ends attached securely to the trees that would be its supports. There is a vast amount of radio literature devoted to getting dipoles into trees. You can find articles with detailed descriptions of how to make a durable connection between coaxial cable and antenna wire. There are thoughtful articles on how to use slingshots, kites, and even small potato-launching cannons ("spud guns") to get lines over recalcitrant trees. Clearly, a little research and forethought make the task a lot easier. But enthusiastic 13 year-olds are not big on research and forethought. They are men of action! They are impatient! They want to get things done, even if the lack of planning and preparation means that the task at hand will take ten times as long as necessary and produce inferior, short-lived results.

My method was as follows:

1) Gather bits of wire. Possible sources include electrical extension cords not currently in use, or the wires for table lamps that were not (at the moment of inspection) actually turned on.

2) Strip insulation off the ends of the wire using the standard insulation stripping tool (teeth). Try to avoid the involvement of your expensive orthadonture apparatus (unless, of course, the sharp edges are found to aid in the insulation stripping).

3) Hastily twist the bits of wire together. Don't solder them together—this would waste time (and besides, you'd need an extension cord for the soldering iron. See step 1.)

4) Quickly measure out the quarter wavelengths using the formula from the radio handbook. A measuring device would be helpful at this point, but being 13, you won't be able to find one, so try to find something that is roughly one foot long: Your dad's size 12 dress shoe is an obvious choice. Be sure to leave said shoe outside when that summer thunderstorm suddenly requires you to go indoors.

5) Attach the antenna wires to the coaxial cable by the same solder-free twist-together technique described in step 3.

6) Find some rocks that can be tied to string and thrown over suitable branches. Spend hours trying to pitch the rock-string apparatus over the selected branches. At this point being 13 has some distinct advantages—just think how silly you will look when you try to do essentially the same thing at age 43!

Of course, many of the strings got tangled and broke, leaving a number of rocks hanging precariously from branches in our backyard, like some sort of weird work of modern art. I think some of them stayed up there for years. Sometimes I'd worry about my dad getting conked on the head by the results of my slapdash dipole construction.

So, as you can see, even the most simple of radio tasks can prove to be difficult for the novice. Hilmar helped when he could, but his time was limited—he was working full time and had his hands full with his own kids. He did come over to the house from time to time to take a look at my evolving station. There were usually some very Germanic expressions of surprise and horror. For safety reasons, for example, all the outside cabinets of all the radio gear in a station should be wired together and then attached to a ground connection—the household plumbing system will do. This prevents you from getting a nasty shock from the gear should something go wrong with the equipment. I, of course, didn't pay enough attention to this requirement—my ground wire was woefully un-Germanic. Instead of the sturdy, thick copper wire that should have been bolted to the DX-40 cabinet and clamped down to the nearest bit of plumbing, I had a flimsy bit of very thin insulated bell wire that was kind of wrapped around one of the screws on the cabinet. From there, it kind of meandered on down to the pipes, where it was again rather pathetically wrapped around the pipe. Poor Hilmar—he really found this kind of thing painful to look at.

UNDERSTANDING: SOME OF THE BASIC EQUATIONS

> *"When I was young, what I call the laboratory was just a place to fiddle around, make radios and gadgets and photocells and whatnot… Well, in that lab I had to solve certain problems. I used to repair radios. I had to, for example, get some resistance to put in line with some voltmeters so it would run in different scales. Things like that. So I began to find these formulas, electrical formulas, and a friend of mine had a book with electrical formulas in it… It had things like, the power is the square of the current times the voltage. The voltage divided by the current is the resistance and all; it had six or seven formulas. It seemed to me that they were all related, they really weren't all independent, that one could come from the other. And so, I began fiddling about and I understood from the algebra that I'd been learning in school how to do it. I realized that mathematics was somehow important in this business."*
>
> From *The Pleasure of Finding Things Out – The Best Short Works of Richard P. Feynman* pp 227-228 Perseus Publishing 1999

As a kid, I guess I had the same kind of "laboratory" that Feynman had, but unfortunately I didn't make the connection between the math I was learning in school and my home experiments with batteries, light bulbs, wires and (later) radios. For me, the "fiddling about" with the equations came much later, and was prompted by an effort to understand how transistor amplifiers work. Most amateurs know that Power = Voltage X Current or P=EI. But to understand how the load for a transistor is determined, you really need to know *why* P=EI. You also need to really know what we mean by power, volts, and amperes.

I had to go back and re-learn some definitions from high school physics:

- **The newton is the unit of force.** One newton will cause one kilogram of mass to accelerate by one meter per second for each second it is applied.

- **Energy is the amount of work done or the capacity to do work.** The joule is the unit of energy. A joule is one newton applied through one meter. Lift an apple above your head and you have expended about one joule of energy. You have also given that apple one joule of potential energy (energy that could be released were you to let it fall).

- **Power (P) is the measure of the *rate* at which work is being done.** The unit is the watt. A watt is one joule per second. The way we use the word "power" in everyday life is similar to the definition of power in physics: We say something is powerful if it can get a lot of work done fast. Move that apple above your head in one second, and that's one watt. Take two seconds to lift it, and your rate of energy use would be ½ watt. Think of watts as joules per second. That 100 watt bulb above your desk is using 100 joules of energy each second.

- **Voltage (E) is a measure of electrical potential.** It describes how much energy is given—or could be given—to a charged object between two points. The volt is the unit. One volt would impart one joule of energy to one coulomb of electrons. A coulomb of electrons is defined as 6.24×10^{18} electrons. That's more than 6 billion billion electrons.

- **Current (I) is a measure of how many electrons pass a given point per second. The unit is the ampere. That's one coulomb per second.**

- **Resistance to current flow (R) is measured in ohms. Different materials have different levels of electrical conductivity. If you place one volt across a material, and one ampere flows through that material, we say that that material has 1 ohm of resistance. This is Ohms law: E = IR**

You can see why in discussing electrical power, we have to include both current (I) and voltage (E). I tells you how many electrons you are working on, and E tells you how much energy you are giving to each.

E=IR or I=E/R or R= E/I

These are really just definitions. For example, we define 1 ohm as the resistance that will cause a voltage difference of 1 volt to appear when 1 ampere of current flows through it. One volt will cause 1 ampere to flow through 1 ohm.

To understand more deeply *why* **P=IE, we need to explore the interaction of newtons, joules, volts, amps, ohms and watts in simple electric circuits.**

Start with a 1 volt battery. Chemicals in the battery have caused there to be an excess of electrons at one terminal, and a shortage of electrons at the other terminal. We say there is a difference in electric potential. One volt means that each coulomb (each group of 6 billion billion electrons) will be given one joule of energy if it is placed between the battery terminals.

Connect a wire to each terminal of the battery, but don't connect the wires to each other. Free electrons in the wires feel a force, but they won't be going anywhere because of the gap between the wires. Think of that gap (which we call an "open" circuit) as being the equivalent of a very, very high value resistor. In mechanics, this would be the equivalent of pushing against a large object, but being unable to get it moving (perhaps because of friction). Voltage: 1 volt, Current: 0 amperes. Energy given to each coulomb: 0 Joules, Power: 0 watts.

Next let's consider what happens when current does flow. We take a 1 ohm resistor and connect it between the terminals of the battery. By definition, 1 volt will cause 1 amp (one coulomb per second) to flow when the resistance is one ohm.

Voltage across the resistor: 1 volt. Current through the resistor: 1 ampere (one coulomb per second). Energy given to each coulomb: 1 joule. Power: 1 watt

Power is defined as the rate at which work is done. P= joules/second

Voltage (E) is defined as the amount of work done on each coulomb of charge: E=joules/coulomb. Current is defined as the numbers of coulombs per second I=coulombs per second. So when we multiply joules/coulomb X coulomb/second, the coulombs cancel out and we are left with joules/second... which is the definition of power.

$$\text{But why does } P = E \times I?$$

$$\text{Definition: } P = \frac{\text{Joules}}{\text{Second}}$$

$$\text{Definition: } E = \frac{\text{Joules}}{\text{Coulomb}}$$

$$\text{Definition: } I = \frac{\text{Coulomb}}{\text{Second}}$$

$$E \times I = \frac{\text{Joules}}{\text{Coulomb}} \times \frac{\text{Coulomb}}{\text{Second}}$$

$$\text{So } E \times I = \frac{\text{Joules}}{\text{Second}} = P$$

Now let's increase the amount of resistance to 2 ohms. Mechanically this would be like increasing the friction on the surface under the large object you are trying to push. Voltage: 1 Volt. Current through the resistor: ½ amp (1/2 coulomb per second). Energy given to each coulomb: still 1 Joule, but now only half as many coulombs are passing by each second. So Power is reduced to ½ watt.

Let's see what happens if we now increase the voltage across our 2 ohm resistor to 2 Volts. Current is now back to 1 amp. Each coulomb is now given 2 joules of energy, and one coulomb is passing by each second. So power is now up to 2 watts. Doubling the voltage has led to a fourfold increase in the rate at which energy is used.

Now let's combine the power equation with Ohm's Law.

Power Equation: P=IE or I=P/E or E=P/I

Ohm's Law: E=IR or I=E/R or R= E/I

In P=IE, substitute IR for E. P=IE becomes P=I(IR) or $P = I^2 R$

This makes sense: If you keep the current the same as you increase the resistance, you must be giving more energy to each coulomb, so the power expended is increased. Imagine a conveyer belt (the wire) carrying a series of boxes (the electrons, one coulomb in each box) up a slope. Imagine that you are at the bottom of the slope, pushing the train of boxes along. If you keep the rate at which the boxes go up the slope constant but then increase the angle of the slope (increase the resistance) you must be increasing the amount of energy expended on and imparted to each box. The rate at which the work is done (P) has increased.

In P=IE, substitute E/R for I. P=IE becomes P= (E/R)E or $P = E^2/R$

This also makes sense. Using the conveyer belt example, imagine keeping the amount of pressure you are exerting on the boxes (E) constant. As the slope of the conveyer belt (R) increases, the rate at which the boxes go up the slope will decrease. The rate at which work is done (P) will decrease. In other words, if you maintain a constant voltage, but increase the resistance in the circuit, the rate at which energy is expended (the power) will decrease—you are giving each coulomb the same amount of energy, but because of the higher resistance you are moving at a slower rate.

These are very basic equations. They are really just a set of inter-related definitions. It is very helpful to understand what's behind them, and to relate them to more familiar mechanical examples from everyday life. This deeper understanding of the basic equations definitely makes it easier to understand how and why power amplifiers work (more on that later). But these aren't the kind of equations that offer deep insights into the physics of radio. Those insights are found in equations like those of Maxwell.

****

With Hilmar's help, my first station was finally assembled, and it was time for me to venture out onto the airwaves. This was very exciting. In my eyes, that first radio contact seemed like it was up there in importance with a pilot's first solo flight. I got quite a bit of help with that first contact, because it was with good-old Hilmar. On July 19, 1973, from across town he was listening for me. In my slow and nervous Morse code, we exchanged signal reports. I was "on the air." I'd made a contact. WN2QHL had wiggled the ether and communicated via electromagnetic waves.

There was no stopping me now. I became obsessed with talking to other stations, obsessed with the distances involved. Ohio! I talked to someone in Ohio! I'd pull out the maps and carefully measure the distance to each station contacted. 487 miles! Amazing!

One of the best things about these contacts was that I was communicating with people who seemed as thrilled as I was about the whole thing. Amateur radio may be unique in what I guess we could call its "culture of CQ." CQ is telegraphic shorthand for "a general call to any station." It is the ham radio way of shouting: "Hey, is there anyone out there who can hear me and wants to talk?" Almost anywhere else in human society, a stranger heard shouting this question would raise eyebrows and stir suspicions—the response would probably be "Well, who are you, why are you seeking a conversation in this way, and what would we talk about?" That's one of the things that makes ham radio special: these kinds of questions are never asked. When he hears a CQ, the other ham knows the answers: The caller is a fellow radio amateur. He's calling CQ because that's what we do (a deeper answer would involve some of the social isolation and geek loneliness issues discussed earlier, but all this would also be understood by other hams). And as for what we would talk about, well we'd start with the standard name, location and signal report. From there we could move onto just about any other topic. Tech stuff is probably the most common subject. (Discussions of politics and religion are strongly discouraged.) Even with all the advanced technology of the internet, even with modern Voice over IP and video technology and the tens of thousands of specialized discussion groups, I don't think there is anything that quite matches ham radio's CQ, and the fraternal response that it provokes.

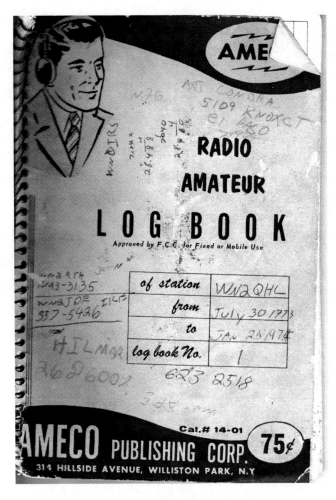

Radio amateurs were required (by Federal Regulation!) to keep log books. Each call or contact was duly recorded. This was serious business. Federal business. There was always the possibility of a knock on the door by officials demanding to see the station log. I still have my log books from this period. Looking at them now, it is a good thing that the Feds never showed up at my door—I imagine they'd be expecting neatness. My first log book came from the AMECO company of Williston Park, N.Y. and sold for seventy five cents. It was in spiral–notebook form and had on the cover the head-and-shoulders drawing of a man in a suit and tie who looked like he'd stepped right out of a 1950's television series. He was a serious, square-jawed, clean-cut fellow. And you knew he was a radio amateur because he had headphones on. The cover indicates that it records the activities of station WN2QHL from July 30, 1973 to January 25, 1974.

Neat, it is not. The margins of the log pages are covered with my frantic efforts to copy down Morse code letters that were coming in far too fast for me. ...TNX BILL NAME IS AL AL IM IN WATERTOWN MAS ?X ML WEST OF BOSTON WN2QHL DE W1TRS K

The allure of the microphone and voice communication was still very strong, but I was finding Morse code to be a lot of fun. You got better at it over time, and it became easier to understand the torrent of dots and dashes that were coming at you. You became more familiar with the peculiar Morse shorthand and traditions that had developed over the years, and this made you more and more part of the fraternity. A fellow ham was addressed as OM (Old Man—this one seems very British). "Good morning" is shortened to GM. "Thanks" becomes TNX. There is a large collection of three letter Q signals that are used to make up for Morse's inevitable letter-by-letter slowness, and to bridge language gaps: "QTH" means "my location is..." "QTH?" means "what is your location?"

This was all very satisfying. I was using a hard won skill, doing something that not everyone could do, and getting better at it each day. And there is something very techno-romantic about the sound of a weak Morse code signal coming in over the airwaves from a distant, exotic location.

My second contact was with a kid named John. His call was WN2RTH and he lived in nearby Nanuet, N.Y. We had met at the club, and the log-book entry indicates that our contact was pre-arranged. John's call sign shows up very frequently in my early log books. The distance between our stations was so short that we didn't need the ionosphere to communicate—we could speak directly via waves that traveled along the ground. So when radio conditions were poor (as was frequently the case) John and I would often find ourselves as the only two stations on the radio bands. And because we weren't into after-school sports, our schedules were sadly similar. Many an afternoon I'd tune across a hopelessly dead 15 meter band, only to find John banging out a forlorn CQ, or he'd tune across and find me doing the same.

I was in front of the radio quite a bit during that first summer "on the air." Most of my contacts were with other novices. We were all limited to special portions of the ham radio frequency bands. These areas formed a kind of electromagnetic playground in which the tolerance level for mistakes and inexperience was very high. Most of my contacts were on the 40 meter band (this corresponds to frequencies of around 7.1 MHz) and the vast majority of my contacts were within about 1000 miles of downtown Congers…

11:45am July 26, 1973: I called CQ on the forty meter band. WN8PNQ responded to my call. Clay was in Columbus, Ohio.

10:45am October 10, 1973: I called W2HAG on 40 meters. Bob in Syracuse.

My log books show 179 Novice Morse Code contacts, almost all of them very similar to these.

There were technical ups and downs. In early February 1974, one of the guys in the club called me on the phone to report that although I was transmitting on the 40 meter band, I also had a signal going out on the 20 meter band. Oh no! A dreaded harmonic! If a transmitter is not properly adjusted or filtered, it can transmit on a multiple of the desired frequency. A cold chill ran down my spine. I almost peered out the window, half expecting to find FCC G-men in dark suits coming up my front steps. I dutifully went off the air until the technical difficulties could be resolved.

Someone in the club must have realized how unbearably frustrating being off-the-air must have been to an enthusiastic 15 year-old. The log shows that I was soon back on the air with a new and far better transmitter. Gone was the little DX-40. Now I was on the air with a huge DX-100. This was the same transmitter favored by those guys on 75 meter AM. This was a real transmitter.

The DX-100 weighs over 90 pounds. It comes in a battleship grey cabinet. The top is a made of a strong mesh-like metal that allows the considerable heat from its many vacuum tubes to escape. Decades later, after kids like me grew up and got nostalgic, transmitters like the DX-100 came back into vogue. By this time tube type rigs had largely been replaced by light and efficient transistorized transceivers from Japan, but as with old Chevys, there was something special about these old-tech transmitters of our youth. Because they were so much heavier than the modern rigs, the old transmitters were all placed in the affectionately named category of "boatanchors." They were heavy enough to anchor a small vessel. "Boatanchors" eventually came to refer to all old tube type amateur radio gear, but legend has it that the term originally applied specifically to the DX-100.

I loved that old transmitter. It had so many big tubes in it that it helped heat the room that I shared with my younger brother. It pulled so much current that the lights in the house would dim and flicker as I pounded out Morse code. In addition to solving my harmonics problem and keeping me out of hot water with the dreaded FCC feds, the DX-100, allowed me for the first time to go international.

Foreign stations are, in ham radio slang (derived from Morse shorthand), referred to as DX (for distance). In a certain sense, DX is what radio is all about. Heinrich Hertz started out with signals that would only reach across his lab. Guglielmo Marconi tinkered away, trying to get his signal to go further and further out across the Italian landscape. Then there were oceans to cross, then signals from space… The whole point of radio is to send your signal as far as you possibly can. So in a certain sense, all radio enthusiasts are DXers. On its first night on the air from Congers, N.Y., that DX-100 gave me my first taste of real DX. The call sign was WA6TLJ/HK6. The WA6 indicated a U.S. station from California. But the suffix HK6 after the slash (little did we know at the time the future importance of /) meant he was operating from Colombia, South America. Wow! DX! I was very excited. I had been recording the times of my contacts using internationally standardized Greenwich

Mean Time (GMT) but this contact was so important that I didn't want to risk miscalculating the time difference, so I recorded the moment in local time. 10:55 pm. I made him send me his mailing address so that we could exchange post cards (QSL cards) and confirm this (for me) momentous achievement.

Looking through the log books from that period brings back some bitter-sweet memories. By this time I was hanging around with four kids that I'd known from elementary school. I considered our little group inseparable, but, well I guess we weren't. In what was in retrospect an early warning of my not being a member of the "in crowd," when the 1973 Easter school break rolled around it was decided that that gang of ours was going to take a trip to Kentucky to visit another fellow from our class who had moved away. But somehow I wasn't selected for the trip.

It's tough when you are fourteen and your entire social network suddenly decamps for Kentucky. I can see log book entries in which I was forlornly trying to pass messages to my friends via ham radio. (This was a holdover from the earliest days of ham radio—there was a network of amateurs that would relay birthday greetings and other messages. The message would bounce around among ham radio stations until it made its way to within the range of a local phone call, at which point it would be delivered.) I was clearly in that awkward border zone between kid-dom and teen-land. I obviously didn't realize that my geeky attempt to reach my friends via ham radio was probably marking me even more clearly as one of the not-cool. But I think this kind of social rejection intensified my involvement in amateur radio. The glow from my rig was like a friendly campfire around which I could always find friends.

Looking back, I wish I'd developed closer ties to fellow radio geeks in the local area. Today, when I read about the teenage technical adventures of people like Steve Wozniak and Bill Gates, I realize that I missed out on a lot by not hanging out with the local friends who I'd met via ham radio. I foolishly tried to fit in with the "cool kids."

My novice year seems to have percolated along uneventfully. Even though a year seems like an eternity to a fourteen year-old, I was acutely aware that my novice license had an expiration date on it. The clock was ticking. It was up or out. I had to pass the test for a higher grade of license, or face the ignominy of being banished into the ranks of failed novices, perhaps doomed to eventually seek comfort in that street gang of the radio world, the Citizens Band. "10-4, Good Buddy," and all that. It was enough to keep a novice up at night.

I was shooting for the General Class FCC license. This would give me access to almost all of the amateur radio frequencies, and at higher power. Most importantly it would allow me to finally plug in a microphone and join in the fun on voice.

Unlike the Novice exam, the General was serious business. It wouldn't be administered by friendly Hilmar and the guys at the club. Oh no, for THE GENERAL you needed to go to the THE FEDERAL BUILDING in New York City. There an actual FCC inspector would test your Morse code speed (you needed a minimum of thirteen words per minute) and give you the technical exam.

I would dutifully tune into W1AW at the appointed hour and struggle to understand code at ever increasing speeds. The backs of the pages in my log books are covered with my frantic scratching down of Morse replays of old *QST* technical articles:

> ...PRODUCES A RAGGED SOUNDIN P LSE IN THE MOBILE RECEIVER... DOES NOT APPEAR WHEN DRIVING ON DIRT ROARS IS A SURE INDICATION THAT TIRE STATIC EXITS...

Preparation for the technical exam had me poring over the ARRL Handbook, trying to really understand the theory that would be covered on the exam. My dad later told me that he would often return home from work after midnight, and find on the dining room table my notebooks and the ever-present ARRL Handbook. Usually the Handbook would be open to the schematic diagram of whatever circuit it was that I was trying to understand. It was all Greek to him—he must have wondered why and how I'd become so deeply hooked on this stuff.

I was on the air just about every day, usually on the 40 meter band. Typically I'd talk to two or three people, almost all of them Americans east of the Mississippi. For many months that contact with the station in Colombia was my only DX contact. His confirmation post card (QSL card) arrived, and that was exciting, but I was after bigger fish. Colombia seemed almost local. Heck, you could drive there! Like Marconi, I wanted to span the mighty Atlantic. I pictured the radio waves from my DX-100 traveling out through the night, bouncing along, skipping between the dark ocean and the ionosphere before finally arriving at the shack of some like-minded member of the radio brotherhood.

By this time I was a subscriber to *QST* magazine, the monthly journal of the American Radio Relay league. *QST* was filled with interesting but for me, incomprehensible, technical articles. The March 1973 issue, for example, features these: "An Inexpensive Time Domain Reflectometer" (that you would build from scratch, of course); "Simple and Efficient Feeds for Parabolic Antennas" (for those going for that "radar on the roof" look). Incomprehensible or not, I read every page of those magazines. Even today, 35 years later, whenever I come across a *QST* from 1973-1975, I recognize and remember the cover. I'm not alone in this madness. Jean Shepherd said that as a teenage ham, he read even the small ads in the back of *QST*. He recalls reading with keen interest about the different widths of grommets available from *QST*'s advertisers (even though he did not at the time know what a grommet was!).

For me, the most intriguing part of the magazine was a monthly column called "How's DX" by Rod Newkirk. This column just reeked of long-distance radio adventure. At the top of the page the title was written on what looked like a scrolled treasure map, and—sure enough—off to the right there was a tall-masted sailing ship (no doubt in some exotic foreign harbor). To the left there appeared a tropical island, with palm trees and—wow!—a thatch-roofed, ocean side radio shack, complete with a dipole antenna (just like mine!) fed at the center with cool-looking and no-doubt homemade open-wire line.

The header was good enough, but the contents of the column were enough to keep a teenage radio fiend absorbed for hours. In the March 1973 issue, Rod begins with a radio-ized version of the Desiderata ("Go placidly amidst the noise and haste… Transmit quietly and clearly, and listen carefully to others…") There followed the usual five pages of brief and mysterious "bulletin board" reports on the doings of our brothers beyond the seas:

"4K1C is said to radiate from Antarctica's Russian Vostok base, further implementing the U.S.S.R 4J-4K-4L prefix blocks… SM2AGD/CE0's impressive Easter Island QSL [card] was well worth the 45 day wait… I'm quite active on 10, 15 and 20 sideband from Kwajalein with a GT-550A and TH3-MK3…"

Vostok Base! Easter Island! Kwajalein! Those were the kinds of places that real hams wanted to talk to. We dreamed of going into the ham shack on a cold winter's night, firing up the old rig, and sitting back to have a friendly chat with good old Vlad down there at Vostok Base. Maybe ask him about the penguins.

Even better than the text were the photographs. The foreign amateurs usually seemed very foreign and exotic. They had names like Miguel and Kiril and Pano. But because they were almost always seated in front of their rigs, there was something definitely familiar about them. They were like us. They were in the brotherhood. Indeed, not even the Cold War was allowed to interfere with communications within the fraternity. We all knew, of course, that Vlad down there in Vostok Base was with the Commies, and might have been planning to incinerate us, but ham radio was neutral territory where all that geopolitical heartache was put aside. How are the penguins today Vlad?

The rigs they were sitting in front of often marked the foreigners in "How's DX?" as especially worthy members of the fraternity. Far more frequently than was the case with U.S. amateurs, the exotic foreigners in "How's DX?" were using homemade equipment. Homebrew in ham talk. There'd be 16 year-old Serge from Kiev sitting calmly in front of a refrigerator-sized rack of gear with more meters and knobs than a Mission Control launch console. The caption would read: "Serge, who just started at Kiev Polytech, designed his own rig and built it entirely from parts scrounged from discarded Russian-made television sets." Serge did not play around with pretty, plug-and-play, commercially manufactured radios. He was not an "appliance operator." He was part of the same radio-building brotherhood that Shepherd had talked about. Like Shep and Stan and Bollis, Serge, you see, was a REAL ham. A ham's ham. A wizard. A builder of rigs. A radio maker.

I wanted to be like Serge. As I studied the technical material for the General class exam, I decided that I would not be satisfied with merely memorizing the material. I wanted to deeply understand the theory. I wanted to really know how the radios worked.

This desire to deeply understand the theory frequently led me down Feynman's "dark alleys" and into FA Wilson's "deep end." But I wasn't the only kid involved in this kind of struggle. In his excellent book *History of QRP in the U.S., 1924 – 1960*, Adrian Weiss, W0RSP, writes:

"But it was very difficult for a boy, working on his own, to grasp how all of this 'radio stuff' actually worked. I recall puzzling for days over a receiver circuit front end tank coil. I understood that, in electric circuits, wires had to be connected, and I couldn't see how the circuit could work if the primary and secondary windings of the tank circuit had no physical connection. At first, I decided that the printer had made an error and left out a line which would show the wire connecting the primaries and secondaries of all the coils. Pleased with my discovery, I dutifully drew connecting lines between the primaries and secondaries of all the coils to correct the printed schematic, and to save the next guy who got the book the trouble of figuring out what I had just corrected, or worse yet, of actually trying to build it the way the schematic showed, and being disappointed when it didn't work. I'd study one section of the Handbook, and go back to another, and eventually, each time I would go back another question would be answered, another aspect of the 'radio magic' would make sense. This took a long time and a great deal of puzzling over the secret mysteries I was attempting to master…"

I struggled on, making mental note of the many areas in which my understanding was less-than-complete, doing my best to teach myself all about various kinds of oscillators (the stage in a radio where radio frequency energy is generated), modulators (where audio, video or other forms of information is added to the signals), amplifiers (there are many, many kinds and configurations) and antennas (seemingly simple, but actually the most mysterious and difficult-to-understand of all). All the while I was banging Morse code into my head at higher and higher speed… with the clock ticking down to the April 27, 1974 expiration of my Novice class FCC Amateur Radio license.

Exam day finally came: Tuesday, March 5, 1974 at the Federal building in downtown Manhattan. My dad drove me into the city, and made sure that I got into the exam room to face my inquisitors.

It has been many years, and my memories of the day are less than clear. I remember the code exam being given in a small narrow room with rows of little schoolroom desk/chairs. I think we did the code first. I knew I had passed—this was a big morale booster. Then it was on to the theory. And I passed that too. Wow, that was a feeling of achievement! I was 15 years old and felt as if I'd just conquered the world.

Now began the long wait for my license. I think they told me that it could take up to six weeks. But I knew that the wheels were now turning in FCC license headquarters in Gettysburgh, Pennsylvania. As I waited for the arrival of my new license and new privileges, some equipment upgrading seemed in order.

I needed a better receiver. In spite of its jeweled movements, the Lafayette HA-600A was not a really serious ham radio receiver. It was designed for casual short wave listening, not for use in two-way communication. It performed fairly well on the 40 meter band, but on the higher frequencies (where most of the exotic DX stations lurked) it was difficult to tune and a bit mechanically unstable—if you bumped into the table, it would shift frequency and you would lose your quarry.

Hilmar had a much better receiver for sale. It was a Drake 2-B. Using money that I'd saved from my newspaper route (The Journal News) and from my evening job as dishwasher (Mayo's Restaurant), I bought the receiver. It was obviously a momentous event for me—the date is marked in my log book: April 11, 1974.

The Drake 2-B revolutionized my radio operations. It was as if I'd been cured of deafness. I could hear stations that I'd never heard before. It was extremely stable—no more sudden frequency shifts. And selective! While with the HA-600A I'd had to listen to several Morse Code signals at the same time, the filters of the 2-B allowed me to focus in on one at a time. This was luxury. And the results were dramatic. Looking in the log, where I'd been previously talking to 3-4 stations per day, on April 13 I had contacts with 26. It appears that I spent most of that day in front of the radio. And that night I had my big DX breakthrough.

I'd dreamed of "crossing the pond." I'd crossed the Caribbean, but that seemed more like a puddle. I wanted to send a signal all the way across the ocean. I pictured my European counterparts in quaint villages, perhaps smoking pipes, tuning in my Morse code salutations. But so far, all I had on the DX scoreboard was one contact with nearby Colombia.

At 10 p.m. on April 13, near the end of my all-day radio marathon, that Drake 2-B pulled in the most exotic radio call sign that I'd ever heard. At first, I couldn't believe my ears. ZL2ACP. ZL! New Zealand! Holy Cow, I'd been waiting to cross the Atlantic, and here was a chance to cross the mighty Pacific! New Zealand was about as far as you could go. Any farther and you were coming around the other side.

I waited for him to finish his conversation with another amateur. I could not possibly have been patient. At the very first opportunity, I gave him a call and—miracle of radio miracles—he responded. His name was Vic. He gave me a report on my signal (559 – not bad) and then—icing on the cake—he told me where he was: Waipawa, New Zealand. Waipawa. That was the coolest, most exotic-sounding location I had ever heard.

I remember bounding up the stairs to tell my parents of my trans-Pacific achievement. It was as if the radio gods had rewarded me for passing the FCC General Class exam. I was now well and truly a DXer.

I was now, of course, deeply hooked, and I yearned for more DX. The next day I spanned the continent by contacting WA6UUR, Jack, in Pasadena, California. But after the New Zealand contact, California seemed almost nearby. Two days later I crossed the Atlantic—I spoke with several Yugoslavian stations. Then, on April 21, 1974, with my General Class license still being processed by the FCC, my Novice license reached its non-renewable one year expiry point, and I was forced off the air. In the log book, I solemnly note "END OF NOVICE OPERATION."

I knew that the arrival of my license was imminent. Every day, I'd be out at the mailbox, waiting for that little U.S. Postal Service jeep. Finally, after what seemed like an eternity, it arrived. The long-awaited federal-looking envelope from the Federal Communication Commission in Gettysburg, Pennsylvania. I ripped open, and saw… OH NO! A brand new TECHNICIAN class Amateur Radio license. My heart sank. I had passed the examination for the GENERAL Class license, but somehow they had screwed it up and issued me a lower class of license. I was really upset.

Technician class was a kind of radio limbo. To get this license you had to pass the same technical exam as the General class licensee, but the Morse code requirement was only five words per minute, not thirteen. So essentially, Technician licensees were guys who had all the needed technical knowledge, but for some reason couldn't pass the Morse code exam. The poor Technicians were banished to the Very High Frequency (VHF) and Ultra High Frequency (UHF) bands. This meant that they were excluded from the High Frequency (HF) bands, the bands where signals would skip off the ionosphere and travel around the world. In their electromagnetic purgatory, the poor Technicians were largely limited to line-of-sight communication. There'd be no HF ionosphere skipping for them. And I'd been unfairly sent into this radio nether-world.

The guys at the club must have understood my disappointment and frustration. One of them immediately loaned me a VHF transceiver (a combination transmitter and receiver) that would at least allow me to stay on the air and finally allow me to make the voice contacts that I'd been waiting to make during my long year of novice-hood. The log shows that within days of the novice license expiry, I was on the air on AM voice on the 144 MHz band. I couldn't go far with this, but I was able to make a few local contacts, finally with a microphone.

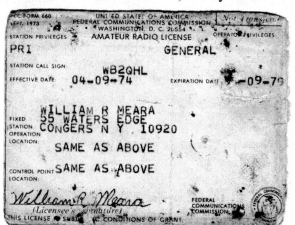

Somehow the FCC snafu was straightened out very quickly. On April 25, 1974 I wrote in the log (with obvious enthusiasm), "END OF TECHNICIAN OPERATION!!" I was out of limbo. WB2QHL, now a fully and properly licensed General class amateur radio station, burst onto the HF airways.

Surprisingly, I was still in Morse code mode. After all that yearning for voice operations, I still did not really have the technology for HF voice operations. Looking back, I wonder why I didn't just take that old DX-100 and put it onto the 75 meter AM frequency that was home to those intrepid youngsters who had had such an influence on me in my short-wave listening days. It may have been that by this time I'd become enthralled with the long distance (DX) potential of the 15 meter and 20 meter bands. Voice operation on these frequencies was almost all in the single sideband (SSB) mode, and I had no SSB transmitting equipment.

Once again the guys in the club came to my rescue. In May or June of 1974 someone either loaned or sold me a Heathkit HW-32A single sideband transceiver. This was a beautiful piece of gear. It was unusual in that it only covered one amateur radio band: 20 meters. But that was fine with me, because over the course of my novice year "20" had become, in my eyes, the promised land. Novices had no privileges on these hallowed frequencies. "20" was only for real hams of General class or above. And "20" was where the DX was.

That HW-32A in many ways embodied the conflicting sentiments that many of us had about commercially made radio gear. On the one hand, we were attracted to it because it looked great. The Heathkit transceivers of this series had these wonderful green cabinets and very aesthetically pleasing front panels. They looked a lot like the legendary and coveted Collins transceivers. In fact, they were known as "the poor man's Collins." We saw pictures of very impressive, neat, NASA-like radio stations filled with these kinds of "radio eye candy" rigs. And we lusted for them.

On the other hand, we knew that real hams—guys like Stan and Bollis and Shepherd and Serge— would look down their noses at this pretty "store-bought gear." We knew that using this stuff would make us vulnerable to the dreaded radio epithet: "APPLIANCE OPERATOR." To the purists, even Heathkits (which had to be assembled) were in this category, especially if you acquired one that had been put together by someone else.

But I soon found out that this particular HW-32A would come to me in a way that would make me at least partially immune to the "appliance operator" tag. When I picked up the rig at the clubhouse, it seemed very light. In the box I found the manual and (YES!) a very cool-looking Heathkit hand microphone. (It was the kind of microphone that the clean-cut, square-jawed guys in the radio magazine ads were always using as they sent typhoon relief to distant villagers, or helped local authorities deal with raging forest fires.) But something seemed to be missing. Where, I asked, was the plug? How did I plug this thing in?

"Oh yea," said the guy who was handing it to me, "it needs a power supply. You can build one yourself. Just get the ARRL handbook and find yourself an old TV."

The off-hand way in which these instructions were issued is a reminder of how much the world has changed since 1974. I sometimes think of that HW-32A when I hear the term "plug-and-play" used in computer shops. Back in 1974, even when using ostensibly "commercial" equipment, ham radio was still very far from "plug and play." I also sometimes think of this episode when I read all of the many warnings that come with any toy sold today; what I was being asked to do would, by today's standards, be considered very dangerous and would probably lead to multiple lawsuits.

You see, the HW-32A was a tube-type HF transceiver. That meant that it needed high voltage. Real high voltage. Not the measly 115 volts of alternating current that comes out of your wall socket, but serious voltage, 800 volts DC. That, my friends, is more than enough to send an intrepid and earnest 15 year-old flying across his basement ham shack, and perhaps into the great beyond.

In a very casual manner, as if it was a trivial, easy-to-do kind of thing, the club members were telling me to:

1. Find an old TV.
2. Strip out the parts that produce the high voltage.
3. Consult the ARRL handbook for an appropriate schematic diagram.
4. Reassemble the TV set parts into a power supply that would succeed in firing up the HW-32A.
5. If possible, avoid electrocution and/or the fiery destruction of the family home.

I had some misgivings, but I guess that new General class license in my pocket made me feel that this was something that I should be able to do. From a discarded TV I began to extract the parts. Most important was the transformer. This would boost the 115 volts supplied by Orange and Rockland

Utilities up to the kilo-volt range (killer-volt seemed like a better designation). I also got the diodes that would convert the high voltage alternating current into pulses of DC. Some big capacitors would smooth out the pulses. There may have been a big coil (choke) in there for additional filtering. The entire big, ugly, and potentially lethal circuit was crammed into and onto a metal chassis about the size of a shoe box.

UNDERSTANDING: EINSTEIN IN THE TRANSFORMER

> *"One ought to be ashamed to make use of the wonders of science embodied in the radio set, while appreciating them as little as a cow appreciates the botanic marvels in the plants she munches."*
>
> —Albert Einstein blasts "Appliance Operators" at the Seventh German Radio Exhibition (Berlin, August 1930)

Even though the guys at the club seemed to treat my power supply project as being only slightly more difficult than changing a light bulb, I soon found that there was a lot of radio mystery in the handful of parts that made up my potentially lethal creation.

Most interesting was that big, ugly, heavy transformer. This device was based on the principal of induction, discovered by Faraday. If you put a changing electric current through a wire, it will *induce* a current in a nearby wire. How it does this gets us directly into some of the deepest mysteries of the universe. Most of the ham radio books don't even mention this, but the truth is that this kind of induction results from Einstein's theory of Special Relativity. I think it is unfortunate that the radio handbooks don't get into the connection between those prosaic transformers and the work of Einstein; I find the connection to be quite beautiful and uplifting. Imagine, we use Einstein's Special Theory of Relativity in our grungy power supplies!

F.A. Wilson, in his book, *From Atoms to Amperes*, explains how this works. Einstein showed that objects in motion undergo length compression when viewed by objects at rest. If you have two copper wires sitting next to each other, there is no electrical force between them. The number of electrons in each atom is balanced by the number of protons. But if you suddenly set the electrons in one of the wires in motion (that's what we call an electric current) this balance is disturbed. Now, according to Einstein's theory, the moving electrons appear to be closer together. Their charges are concentrated in a smaller area. The protons are not moving, so there is no concentration there. Now there is an electrical imbalance, and electrons in the nearby wire will get a tug—there will be current in that second wire. If you make sure that the motion of the electrons in the first wire is constantly changing (as it is in with alternating current) you will get a corresponding to-and-fro movement in the second wire.

Now, if you wind the wires into coils and put them in close proximity, the amount of energy transferred will be increased. By varying the ratio of the number of turns in the two coils, you can vary the voltage coming out of the transformer. In the case of my project, the ratio was such that the 115 volts going in from the wall socket was transformed up to around 800 volts. Special Relativity was putting my young life in danger.

It is also important to think about "self inductance." Remember that a current through a wire will create a magnetic field around it. And remember that a changing magnetic field will induce a changing electric field in nearby wires (and, indeed, in space). Think about what would happen if you suddenly stopped the current through the wire. The magnetic field around the wire would go through a sudden change. The current that had been causing it is gone. So the field is going to change. Sometimes the change is described as "the field collapsing." But the important thing is that it is changing. This changing magnetic field produces an electric field that tries to keep the current in the wire going. You can think of inductance as a kind of electrical inertia. This inertia—this inductance—can be used in circuits to present high impedance to high frequency signals, while at the same time presenting low impedance to signals of lower frequencies.

In an old tube-type transmitter I found one of my favorite applications of this ability to separate high frequency and low frequency signals. In the final output circuit of my Hallicrafters HT-37 I noticed that there was a fairly robust coil connecting the antenna terminal to ground. I wondered what that was all about. Why would you want to ground your antenna terminal? I did some research and found out that this coil was a safety device. In a rig like this, very high DC voltages are present in the final amplifier stage. A capacitor is supposed to block this high DC voltage and keep it in the amplifier stage while allowing the radio frequency energy to go out to the antenna. But if that capacitor were to fail, you'd have 1000 volts DC on the antenna wires. That's a scary thought. The coil from the antenna connector to ground is a fail-safe device. When things are working normally, that coil will look to the RF output like a very high impedance (almost an open circuit)—the RF will proceed out to the antenna. But if for any reason high voltage DC makes it to that antenna connector, that coil will appear to the DC to be a very low impedance path to ground. Lots of current will flow, and the fuses in the power supply will blow, shutting the whole transmitter down. Einstein's Special Relativity will save the life of Johnny Novice working on his antenna up on the roof.

I had problems understanding how transformers worked. For me, one of the most difficult concepts to grasp was how the current in the primary coil is determined by the current in the secondary coil. For example, by changing the value of a resistor placed across the secondary, you would change the amount of current flowing in the primary. But why? But how?

For a long time I struggled with textbook "explanations" that were really just repetitions of equations. I wanted a deeper, more intuitive level of understanding. I wanted an understanding that was more Faraday than Maxwell. I think this is one area in which I was not rescued by some bit of well-written technical prose—on this one I had to struggle on my own to eventually "see it."

It is all really about "self-inductance" and "back emf." When that changing current hits the primary coil, it generates a rising magnetic field in that same coil that opposes the current flow—it creates a back emf. But wait. In a transformer there is another coil involved. In this coil that rising magnetic field from the primary also creates an emf that will cause current to flow in the secondary circuit. And that current in the secondary coil will in turn have an effect on the primary. Here is the key: That induced current in the secondary will be in the opposite direction from that of the input signal. And that current in the secondary will set up its own magnetic field, and this magnetic field will tend to negate the effect of the "back emf" in the primary. So, the more current that flows in the secondary, the less "back emf" there is in the primary, and the more current flows in the primary. That's how the amount of resistance across the secondary affects the amount of current in the primary.

What determines the amount of voltage in the secondary? For me this was easier to grasp. It was set by the turns ratio of the transformer. Start out imagining a transformer with one turn in the primary and one turn in the secondary. Assume that one volt AC across the primary results in one volt AC across the secondary turn. Now add an identical turn within the same magnetic field. Now you'd have 2 volts AC in the secondary. And so on, and so on. Soon you are at the 800 volt level that threatened to blast me and my HW-32A into the land of the silent keys.

This may strike some of you as being to too good to be true. It may smell of cold fusion and perpetual motion machines. After all, we just took two bits of wire and ended up with lots more energy, right? Wrong. More voltage yes, but not more energy. Transformers do not create energy. Power in must equal power out (a bit less actually, when you consider losses). Consider the TV transformer that I used on my power supply project. With 115 volts on the primary, it produced 800 volts on the secondary. If I put a resistance of 1600 ohms across the secondary, by ohms law a current of .5 amps would flow. That would be 400 watts. But guess what? That .5 amp current flowing in the secondary would reduce the back emf in the primary to the point where 3.478 amps would flow in the primary. 115V X 3.478 amps = 400 watts (again, I'm ignoring transformer losses here).

Transformers can do all kinds of other amazing tricks in radio. Among the most useful for us:

- **Separating the AC and the DC.** If you have an AC signal superimposed on a DC current (this often happens in amplifier circuits) a transformer will allow you to separate out the two. Run the combined signal through the primary; in the secondary you'll only have the AC. (Go back to Faraday and Maxwell to see why. "Changing field" is the key term.)

- **Changing impedance levels.** Sometimes an amplifier stage needs to work into a certain level of impedance. But perhaps the speaker or antenna you wish to use is of a different impedance. By using a transformer of the appropriate turns ratio (and the principles outlined above) you can make that speaker or antenna to appear to be the value you need. There are many mechanical analogies for this. Perhaps the best is that of the gear system on a bicycle: Think of the gear connected directly to the pedals as the primary coil, and the gears connected to the back wheel as the secondary. That hill you are going up is presenting a certain impedance to your leg muscles. But as it gets steeper, the impedance reaches a point where you can't pedal anymore. So you shift gears. That's like changing the turns ratio on a transformer. Now you can continue pedaling. Now your legs are being presented with a more suitable impedance.

Some words of caution from Isaac Asimov's *Understanding Physics*:

"Despite the fact that magnetic lines of force have no physical existence, it is often convenient to think of them in a literal fashion and to use them to explain the behavior of objects within a magnetic field. In doing so we are using a model—that is a representation of the universe that is not real, but which aids thinking. Scientists use models that are extremely helpful. The danger is that there is always the temptation to assume, carelessly, that the models are real, so they may be carried beyond the scope of validity. There may also arise an unconscious resistance to any change required by increasing knowledge that cannot be made to fit the model."

****

It took me a while, but I got the power supply to work. The rig didn't actually explode or burst into flames, and I considered this to be something of an achievement. (Later, I would learn that in radio-electronics parlance, my new power supply had "passed the smoke test.") My memories of this project include lots of painful images of frustrating technical problems. I had thought that I never really put this rig on the air, but a look at my log book shows that I did, and with notable success. The first contact took place on June 11, 1974. I was 15 years old. School was out, and radio (among other things) was in! I made dozens and dozens of contacts with that Heathkit. Most were with other U.S. stations, but I was also regularly talking to people in South America and Europe.

Looking back on that summer of semi-homebrew radio, I think I should have taken more pride in the fact that I'd put that power supply together myself. Other than the antennas, that was my first real piece of homemade radio equipment. But I guess because it was feeding that very commercial-looking

Heathkit rig, I didn't feel that my power supply project put me in the same league as Serge and the radio wizards.

My HW-32A operations lasted until November 9, 1974. Then I went off the air for more than four months. I guess this long hiatus from ham radio was due to a combination of technical and hormonal factors. I think problems cropped up with my homebrew power supply. I don't think the voltage output was as well regulated as it needed to have been, and this might have caused problems with my signal. Solving a technical problem like this requires focused attention, but like most 15 year-olds, I was far from focused. I was on the typical teenage hormonal roller coaster. I think the drop-off in log book entries was a clear sign of the intrusion of other interests that almost always beset teenage radio amateurs. You just start getting distracted by girls, and by all the things you have to do to pursue them.

Of course, it is not a sudden change, and you don't completely lose your other interests. And as I'm sure is a common occurrence among radio fiends, lack of success in the social arena frequently sent me back to the warmth of the vacuum tube-heated ham shack.

On the technical side, the Crystal Radio Club had come to the rescue yet again. One of the members was selling a transmitter that would go very well with the Drake 2-B that I bought from Hilmar. It was a Hallicrafters HT-37. Like the HW-32A, it was a single sideband rig, but it had the big advantage of covering all of the high frequency bands. I made the investment. I guess it was a good move, because, like the 2-B, I still have that old Hallicrafters transmitter.

The HT-37 went on the air on March 15, 1975. Unlike the HW-32A, this rig could also transmit Morse code, and my first few contacts were a no-doubt nostalgia filled return to that ancient mode. But soon I was back on the microphone. The Hallicrafters rig caused a big up-tick in activity. I was on the air a lot during that spring.

But as my 16th summer approached, the entries in the log start to become more sporadic. By now a driver's license seemed more important than my ham radio license. And in perhaps the saddest indication that my priorities were shifting, I was now using my very official and adult-looking General Class FCC license in efforts to get into bars while still underage. I was using the FCC license as fake ID. I hang my head in shame.

I was sliding into the debauchery of teen life in 1970s America, but I had not completely lost interest in the radio arts. I can tell from the log book and from a collection of old QSL cards (that my mother saved) that I was on the air from time to time. And somewhere in this period, I got involved in a homebrew project that was more ambitious than the power supply effort.

My use of commercial gear had left me with a nagging sense of incompleteness. Even though I had successfully melted solder and braved high voltage with the HW-32A power supply, deep down I knew that I was essentially an appliance operator. And I didn't like that. I wanted to be a real ham, a homebrewer, a builder of radios.

Through the ham radio magazines, or perhaps through conversations at the Crystal Radio Club, I'd gotten the idea that the real challenge in radio construction was not with transmitters, but with receivers. This notion probably came from the 1930s and '40s, when most hams built their own simple transmitters, but used them with commercially made receivers. Building a transmitter was seen as a routine, relatively simple task, not much more difficult than my power supply project. But building a useful receiver, well that was the mark of a real radio wizard.

In one of the radio magazines (it may have been *73*) I came across a circuit that seemed like just the thing for me. All I remember about it was that it was varactor tuned, and solid state (i.e., it used transistors, not tubes). A varactor is a type of diode designed to vary in capacitance as the voltage across it changes. I decided to use this circuit to lift myself out of the ranks of "appliance operators." I felt that if I was able to build a real ham radio receiver, I would be able to rightly claim a place among the ranks of Serge, Stan, Bollis and the other radio wizards.

Unfortunately, I didn't really know what I was doing, and I bit off more than I could chew. Building a receiver is an ambitious project—you should really have some more construction experience before jumping into a project like this. My marginally successful power supply construction didn't give me enough background. And I did not really understand how the circuit worked before I started trying to put it together.

This was not a kit. I had to make multiple visits to a local Radio Shack store to buy the individual parts. The article recommended building the receiver on a printed circuit board. For this, I would have to get a piece of copper-clad board and then, using chemicals, etch onto it the desired wiring pattern.

When you look at a the schematic diagram of a complex piece of electronics, if you understand the circuit you will see not just a jumble of connected components, but a series of stages, each designed to do something specific to the signal. Different stages will generate, amplify, mix, filter, or modulate the signal. And when you build this circuit, you should build it in stages, (and test each one) even if all the stages end up on the same circuit board.

Because I didn't really understand the circuit, I DID see it as just a jumble of parts, so I probably laid out the circuit board pattern in a very haphazard manner. This is not good. By doing this, you will almost inevitably be putting the output of amplifier circuits too close to their own inputs, and you will set them into undesired oscillations; they will be like little electronic dogs chasing their tails, and you'll get squeals like the ones you hear when a microphone gets a bit too close to the speaker in a public address system. To make matters worse, because you have created such a jumble, you won't be able to easily troubleshoot the circuit.

This receiver was a simple direct conversion design. RF energy from the antenna was amplified, and passed on to a mixer stage where it was combined with a signal from an oscillator stage operating at the receive frequency. When the two signals were mixed, the audio energy was, in effect, separated from the RF signal, and passed to the audio amplifiers.

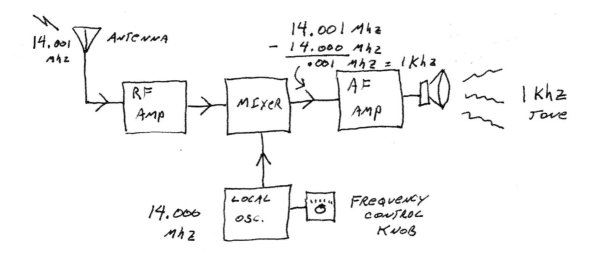

BLOCK DIAGRAM OF A DIRECT CONVERSION RECEIVER

I spent a long time building that receiver, but I couldn't get it to work. Because I didn't really understand the circuit, I didn't even know how to determine which stage or stages of the device were not functioning. Unlike young Adrian Weiss, it never occurred to me that there might have been a defect in the schematic diagram. For some reason, I didn't think to turn to the radio club guys for help, and in pre-internet days there were no USENET groups or mailing lists to turn to for assistance.

I know now that I came very close. I could hear some static in the headphones, so the audio stages were working. And I noticed that when I had the receiver close to my HT-37 transmitter, I could tune in signals on the receiver only when I was generating RF with the HT-37. Looking back, the problem is now obvious: The local oscillator stage was not oscillating. When I turned on that HT-37, some of the RF energy was getting into my little homebrew wanna-be receiver, and was doing the job that the oscillator stage should have been doing. Had I realized what was going wrong, I could have focused my attention on that one problematic stage, and I could have gotten that receiver working. But patience and persistence are not exactly the hallmarks of adolescence…

(I want to take another shot at building this receiver. If anyone can find the article, please send it to me. It came out in the early 1970's in one of the ham or electronic magazines and the distinctive characteristic was that it was "varactor tuned.")

UNDERSTANDING: SEMICONDUCTORS

> *"These semiconductors have long possessed a special interest for electronic researchers. The important thing about them is that the number of current carrying electrons in them can be controlled."* **From the September 1948 article in** *Scientific American* **that heralded the solid state revolution.**

Those little transistors and diodes in my homebrew receiver had a lot of mystery wrapped up inside them, and I wasn't comfortable with that mystery. I was scornful of hams who didn't understand the circuitry in their fancy commercial rigs. For them, their transceivers were really just black boxes. But truth-be-told those transistors and diodes were, for me, just black boxes of a smaller scale. I wanted to understand how those tiny devices worked.

The word semiconductor is a good place to start. Silicon lives in the gray area between conductors and insulators. Usually it is an insulator, but if you heat it up and shake loose some of the electrons, it starts to behave more like a conductor. The trick is to find some way to control or vary the extent to which it is shifting from insulator to conductor. That would be a useful trick indeed.

The conductive properties of the silicon crystals can be varied by chemical doping. If you insert into the crystal lattice some atoms with some "extra" valence electrons, the material becomes more conductive. That kind of doping is easy to understand—the extra electrons just make the silicon more like copper, and less like ordinary silicon. This kind of doping creates what is called "N" (negative) type material.

Electric current flows through N type material in much the same way as it flows through copper wire: The free electrons are normally in random motion. When an electric field (perhaps from a battery) is placed across the material, the motion becomes less random and electrons move away from the negative battery terminal and toward the positive terminal. When an electron exits the semiconductor and goes into the wire to the positive terminal, it leaves behind an atom (that donor atom from the doping process) that "needs" another electron; it leaves behind a positive ion. That positive ion soon acquires a replacement electron coming in from the direction of the negative terminal. At the negative terminal, electrons from the battery find themselves near positive ions in the semiconductor material (free electrons from these ions have moved toward the positive terminal). Electrons from the negative battery terminal jump into the material, pulled in by the positive ion. Electrons exit the material at the positive terminal as other electrons enter on the negative side. Current flows.

There is another way of doping the silicon crystal to make it more conductive. This method is more difficult to understand, and it is here that I frequently got confused. Instead of adding atoms to the mix with extra electrons, you can add atoms that are one electron *short* of the number needed to fill up the crystal's lattice structure. These atoms leave "holes" in the silicon's crystal matrix. This creates P (positive) type material. I found holes—and especially hole movement—to be a difficult concept. Apparently I am not alone in this. On the back cover of *From Atoms to Amperes* F.A. Wilson asks, "Have you ever... thought the idea of holes in semiconductors was a bit much?" Indeed I did F.A.!

Different books present these holes in different ways. For me, the best way to think about this is to first consider the un-doped, hole-free semiconductor material. Each atom in the lattice is tightly bound to the others around it. There are no free electrons floating about. It is more of an insulator than a conductor. Now if you dope it by adding to the crystal structure atoms that in their outer electron shell are one electron *short* of the number needed to fill the crystal structure, there will be holes in this crystal structure. In a certain sense you have made the material more "porous" for electrons.

Perhaps in an effort to treat the two kinds of charge carriers (electrons and holes) equally, some of the books I read used the term "hole flow." I found this very confusing. I had images of protons flowing through the silicon (that doesn't happen—if it did, we'd be talking about protonics instead of electronics) or of entire atoms moving through the matrix (wrong again). Keep this in mind: The only things that really move are the electrons. But with a combination of the right kind of doping and the influence of an electric field, you can move the *electrons* in P type material in such a way as to make it *seem* that the holes are moving around. But in reality, only the electrons move.

Let's look at what happens in the P type material when we put an electric field across it. At this point we just have an electric field—we are not yet providing a circuit path that would allow an actual flow of electrons into and out of the material. At this point most books will start talking about holes moving away from the positive terminal, and bunching up near the negative electrode. But *hole* on a second! What's actually moving? Protons? No. Entire atoms? No way. Remember, what moves in electronics is electrons.

In any silicon crystal, a certain number of electrons are constantly being knocked out of their matrix positions by incoming photons (from heat and light). Normally, these electrons would quickly find their way back into the crystal matrix, pretty much where they came from. But the introduction of an electric field will cause them to move in a certain direction (away from the negative electrode and toward the positive). In the P type material the holes from the doping will give these electrons somewhere to go. Electrons from within the material will move toward the positive side of the field, filling in the holes near that terminal. These electrons are loosely held in these holes, because the doping atoms don't need them. Back in the area in which they originated, these departed electrons will have left behind new holes in the lattice near the negative side of the field. And these atoms will be eagerly looking for electrons to fill these new holes—atoms there will now find themselves short of an electron.

Considering the situation before and after the introduction of the electric field, it will look like the holes have moved up to the negative electrode. But now we know what has really happened: the holes haven't gone anywhere, but the electrons in the material have shifted toward the positive terminal, filling holes there and leaving new holes behind in the crystal lattice. It might look as if holes have moved, but in fact only electrons moved.

Let's now throw in an analogy from the world of diplomacy. Imagine a long dinner table. One person sits in each chair. The chair is the atom, the table is the crystal matrix, the people are the electrons. An interesting, *attractive* person begins to speak at one end of the table, a boring, *repulsive* person starts to drone on at the other end—everyone leans towards the interesting person, but because there are no extra chairs, all they can do is lean. The interesting person's influence is similar to that of the positive side of the electric field; the boring person is the negative side. Now, let's add some P-type doping to the table. Replace some of the chair/person combinations with empty chairs. These are the equivalents of acceptor atoms, P-type doping. Now, when that interesting person at the end of the table beings to talk and when the bore at the other end begins to drone on, people at the table begin to switch from their original seats to fill the empty seats closer to the fascinating person. I suppose people who write about hole flow would say that the empty chairs moved away from the interesting person. That might be a diplomatic way of putting it, but let's face it, that's not really what happened!

Now let's consider how electrons actually flow through P type material: Battery terminals are connected through wires to the material. This provides both an electric field and a circuit path. Under the influence of the field, electrons in the material that have been knocked out of their positions in the lattice by heat or light begin to move from hole to hole toward the positive battery terminal, leaving new holes behind. The positive battery terminal has a shortage of electrons (that's what we mean when we say it is positive) and so electrons begin to flow out of the material and into the wire, heading towards the positive terminal. Back at the wire from the negative battery terminal, the excess electrons in that wire (again, that's what we mean when we say that that battery terminal is negative) find themselves close to new holes in the valence layer left behind by the electrons that departed en route to the positive terminal. So electrons from the negative terminal jump into these new holes and soon begin to make their own way through the porous P type semiconductor.

The key point here is that the resistance of semiconductor material depends on the number of holes (in the case of P type material) or excess electrons (in the case of N type material). If some way could be found to vary the number of these "carriers" in the material, that would be very useful for amplification. Our ability to control the concentration of holes and electrons in N and P type material was an important part of the effort to gain this kind of control, one of the keys that permitted the development of transistor amplifiers like those in my little homebrew receiver.

****

CHAPTER 2
OFF-THE-AIR
AMATEUR RADIO GOES INTO HIBERNATION

"This young man is a high-school graduate. He works for a living, is self-supporting, unmarried and is employed in a technical trade. He is quite well-liked in his community, respected for his knowledge of radio with the respect due an expert. He is somewhat lax in fulfilling his social obligations, not through lack of inclination, but because of lack of time. He'll probably be married soon, and then there will be a hiatus in his amateur career, although he will eventually return to the game. He has been interested in radio for several years, a licensed amateur for nearly three. Such is the typical radio amateur of 1936."

From 200 Meters and Down – The Story of Amateur Radio By Clinton B. DeSoto, published in 1936 by the Amateur Radio Relay League, Inc. West Hartford, Ct. page 2

It is now very fashionable for kids to take a year "off" between high school and college. In the UK this is referred to as a "gap year" and even the Royal Princes have followed the fashion. I guess I was a bit ahead of the curve on this one, because way back in 1976, I decided to do something different right after high school.

Bill Orosz was my math teacher. He was also a Captain in the local Army National Guard unit, which happened to be a Signal Corps company. Through him I learned that if I joined the unit, they would send me away for about one year of active duty for training in the Army. After the rigors of Basic Training (which appealed to my teenage sense of adventure) I'd be sent to the Army's Signal School at Ft. Gordon, Georgia, where I'd be trained as a Multi-channel Communications Equipment Repairman, a.k.a., 31L.

This all seemed to fit perfectly with my plans. The timing was good—the training would fill the gap year very nicely. There would be adventure, and electronics. I'd come home with some money saved and the National Guard weekend drills that would follow would become a source of some extra money for college. I signed up and off I went.

A description of Basic Training does not belong in this book, because there was nothing even remotely related to electronics in it. Maxwell's equations didn't help a bit. But I got through it, and then it was off to Signal School.

Soldiers who were destined for the long repair courses were first put through the relatively short "operator" courses in which they learned how to use the gear that they would be repairing. In my case, I went through the 31Mike course—the Multi-channel Communications Equipment Operator Course. This too turned out to have very little to do with electronics. We were taught (trained really) on how to put up microwave antennas (sadly, this involved just assembling olive drab kits—there was no throwing of wires into trees) and to operate the generators that would power our hill-top microwave relay stations. We were also trained on how to operate the equipment inside the communications vans, but we were taught almost nothing about what went on inside the equipment—nothing really about the theory of radio. This was definitely "black box radio." In many ways it was the ultimate "appliance operator" course. 31Mike was painfully easy, and quite boring. The Army, of course, did its best to make us miserable. Most of our time out of the classroom had us involved in maniacal efforts to pick up the cigarette butts that other soldiers had thrown down the day before. There was also a fanatical campaign to keep the barracks clean to almost surgical standards.

After about 6 weeks it was on to 31Lima. This was the repair course. It was obvious that 31L had required higher scores on the Army aptitude tests than 31M—there were far fewer stupid fights, less drunkenness, fewer arrests, and the level of conversation among the students was definitely at a higher level.

In this course, there was a lot of electronics theory. They wanted us to understand how the complex Pulse Code Modulation—Time Division Multiplexing system worked. So by day we were deep into the digital circuitry. But as soon as the school day ended, it was back into the Army, i.e. picking up cigarette butts and polishing latrines. This wasn't really a lot of fun.

31L was a "self-paced" course. It was broken up into modules, and as soon as you finished one, you moved on to the next. Anxious to end my career as a cigarette butt collector, I decided to move through 31L as fast as I could. The Army provided an additional incentive: Whoever finished first was the "honor graduate" and would win a promotion.

31L was all about troubleshooting—about finding the fault that was causing the complex multi-channel system to fail. Our main tools were the manuals that came with the equipment, and the oscilloscope. With freedom from Army harassment awaiting at the end of the course, I became a very fast troubleshooter.

I suppose it was a good course, and I guess I learned a lot, but there was something important missing at that Signal School. Somehow the Army had managed to squeeze all of the joy and wonder out of electronics and radio. I guess this should come as no surprise—this was, after all, the Army, an organization designed to kill people. They seemed to apply to radio and electronics the same process of de-humanizing routinization that was used to teach people to operate an M-16 or a 155mm howitzer. There is not a lot of room for joy or wonder in that world.

In what little spare time that I had, I tried to reach out to the radio amateurs in the area around Fort Gordon, but this didn't really work out. Most of them were associated with the Signal School, and I got the feeling that they just couldn't see me as a fellow ham—in their eyes, I was a "trainee"—just another of the thousands of anonymous crew-cut teenagers who passed through their town, just another cog in the green machine.

UNDERSTANDING: JUNCTIONS AND DIODES

I resisted the Army's efforts to squeeze the fun out of electronics. During some of the few hours each day that I wasn't on "police call" (picking up cigarette butts) or in a "GI party" (cleaning the barracks), I went to the Signal School's Learning Center and took extra courses on electronic theory. I was spending each day troubleshooting solid-state circuitry. In the Learning Center I tried to understand how all those transistors and diodes actually worked.

Once you understand how doping affects the conductivity of P and N type material, the next challenge is to understand the junction, the area in which a piece of P doped semiconductor is fused with a piece of N type semiconductor. Apart, each bit of doped crystal will conduct electricity, each in its own different way. But if you meld them together, some of the excess electrons in the N material will fill some of the lattice holes in the P type material, and you will, in effect, end up with a zone of un-doped silicon, a no-man's land between the forces of N and the army of P, a de-militarized zone devoid of charge carriers. (This zone will exist only near the junction—you might think that ALL of the P material holes could be filled by excess electrons in

the N material, but this does not happen, because the recombination process creates—in effect—walls of ions that limit the extent of this charge carrier neutralization.)

Recall that un-doped silicon does not conduct electricity very well. At the junction you now have a zone that has been depleted of its doping, depleted of the ingredients that permit electricity to flow, depleted of "charge carriers." You have a depletion zone. This forms a barrier of high resistance between the low resistance N and P layers adjacent to it.

Anything that increases the numbers of charge carriers in this depletion zone will decrease its resistance. For example, simply shining a laser or a bright light on the junction, or exposing it to heat, will increase its conductivity by shaking loose electrons and creating charge carriers. Keep that image in mind—it will help you understand transistors.

Remember how we earlier described how an electrical field could be used to move electrons and holes in doped semiconductor material? Well, in a diode or transistor electric fields are used to move electrons and holes in ways that affect conductivity in the depletion zones.

Take your fused combination of N and P material. Put an electric field across it. First try connecting the positive battery terminal to the N material, and the negative battery terminal to the P material. What do you think would happen? In the N type material, free electrons (from the doping) would be attracted to the positive charge of the battery. They would move away from the junction. In the P type material, free electrons (generated by heat) would, upon leaving their "home atoms" move away from the big negative charge on the terminal and would be filling holes close to the junction, leaving behind new holes, gaps in the crystal lattice closer to the electrode. Some (but not me!) would describe this as holes moving away from the junction. Step back and look at what has happened. The depletion zone has grown larger. The resistance between the two conductive zones has increased.

Reverse the polarity of the electrodes and watch what happens. The negative battery terminal goes to the N type material. Free electrons from the doping atoms are pushed toward the junction. The positive terminal of the battery is connected to the P type material. Free electrons in the P type material (those generated by heat) are pulled toward the electrode. Holes from the doping process are left vacant. Holes generated by heat are also left vacant. The depletion zone is made thinner. If the battery's field is above a certain strength, the depletion zone will be eliminated entirely. The crystal will then be suffused throughout with charge carrying electrons and holes. Current will flow.

The combination of N and P material creates a kind of one way street for electricity. This in itself is very useful, and allows us to do things like converting alternating current (AC) into direct current (DC), or converting AC radio frequency signals into varying DC audio signals that we can hear in headphones.

Here are a couple of practical illustrations of how the one-way street provided by diodes can be useful in radio work:

One of the best ways to destroy a solid state rig is to reverse the polarity of the power supply or battery. This is a mistake that is very easy to make, with disastrous consequences for the circuitry—smoke will definitely be released! But you can use a simple diode to make your rig idiot-proof: Put a diode from the DC power input jack to ground. Make sure it is reverse biased (i.e., make sure it is wired so that it will NOT conduct when the battery is connected properly). When you have the battery hooked up properly it will just sit there, conducting almost no current, appearing to be an open circuit. But during that camping trip when it is dark and you're tired and you get the polarity backwards, that same diode turns into a short circuit. Lots of current will flow from the DC input jack, directly to ground (not through your precious homebrew circuitry). The fuse in the power supply or in the cables to the battery will blow immediately, shutting the whole thing down, and protecting your carefully constructed circuitry. Pretty clever, don't you think?

Here's another simple application: What is commonly called "static electricity" can also fry electronics. That annoying spark that jolts your fingertips on dry winter days represents tens of thousands of volts. That voltage can easily puncture the metal oxide in the field-effect transistors that sit at the front ends of many receivers. These sensitive devices can be protected by simple diodes. Wire two of them in parallel, with different polarity orientations (one with the arrow up, the other with the arrow down). Connect this combination between the antenna jack of the receiver and ground. The radio signals coming in from the antenna will not be affected; the diodes need a voltage of about .6 volts (forward) to overcome the voltage wall set up by the ions when the depletion zone formed—the shortwave signals you are trying to receive will never reach this level, so receiver performance will be unaffected. But when you pull that woolen sweater off and then grab the antenna of the shortwave radio, all the charge on your hand (that will GREATLY exceed .6 volts) will flow through one of those diodes to ground. Neat, eh? Three cheers for the humble diode!

These direct applications are very useful, but the most important thing about these diodes and their junctions is that they provide a way for us to control and vary the number of charge carriers in a piece of silicon crystal, specifically in the depletion zone discussed earlier. That's the kind of control that can be used to make an amplifier. That's the kind of control that led to the transistor revolution that reshaped the world. More on this later...

****

I'd been accepted into the Electrical Engineering program at Manhattan College. I'd applied to that school mostly because my grandfather and two of my uncles had gone there. The plan was for me to start at Manhattan after completing my year of military training. But after a year of the Army's version of electronics, I was kind of turned off to it. My dad (who had witnessed my struggles with the multiplication tables and who had doubts about my math abilities) wasn't sure the EE program was for me. So, just a few weeks before the start of college, I wrote and asked to be switched from Electrical Engineering to the School of Arts and Sciences. I became an International Studies major.

This might seem like an odd choice for a techie geek, but I guess all those contacts with DX stations had stirred an interest in foreign places. I hoped that I was getting onto a career path that would send me (and not just my signals) to those foreign places I'd been talking to.

My interest in ham radio went into deep hibernation during my college years. My log books show very few entries during this period. My stack of QSL cards indicate that I was occasionally on the air, but it is clear that amateur radio was now on a back burner.

The hibernation became even deeper after graduation. In a very common pattern, the demands of a new career and frequent moves caused amateur radio to be put aside. For a few years I was back in the Army—I was spending a lot of time in Central America and Panama. My ham radio activity was very limited and sporadic, and was usually related to my work. In an effort to get ready for Special Forces school, for example, I would load up a rucksack and head out into the woods with map and compass; an old Yaesu Memorizer 2 meter transceiver would go into the rucksack—it helped me load up the pack, and I figured it would be useful to be able to call for help if I broke my leg while hiking. Once, on a training exercise in North Carolina, I tuned the Special Forces transceiver we were carrying to the 40 meter ham band. Using Morse code, I found a fellow who was willing to send greetings to my parents (we'd been in the woods for a while and I knew they'd be glad to hear from me). I carried a little Sony short-wave receiver with me and did some listening from time to time. I once attended a meeting of the Panama amateur radio club—they were very nice to me, but I never had enough time to really become part of the club. My Drake 2-B and the Hallicrafters HT-37 spent most of this period in a storage locker in Fayetteville, N.C.

After a few years in the Army, I joined the Foreign Service—America's diplomatic corps. As part of the application process candidates were asked to write a brief autobiography in which they would explain what had caused them to seek a career in diplomacy. In mine, I said that my interest in foreign places had been stimulated by ham radio. Amazingly enough, they still gave me the job!

For most people, leaving military service means returning to the USA and settling down. But that's not what happened to me. In my case, the foreign fun had just begun. I was sent right back to Central America, back to Honduras. These were chaotic times, and I had very little time for hobbies. But at one point I did take the Drake 2-B out of its box. The tuning mechanism had become very difficult to turn. I found an electronic repair shop in Tegucigalpa that claimed it could fix it. I had my doubts, and found myself feeling very nervous about leaving that old receiver in that strange foreign shop. You get attached to these rigs. I retrieved it a few days later, essentially in the same state that I'd left it.

My duties in Honduras involved frequent travel to insurgent base camps along the Nicaraguan border (the details of these journeys go way beyond the scope of this book, but those who want to learn more about this should refer to *Contra Cross* by the author). Again, there wasn't much time for amateur radio, but I did always carry with me a small short-wave broadcast receiver—as I'd drift off to sleep in the rebel base camps, I'd pull out that little receiver and tune across the short-wave frequency bands. I suppose it was something of an escape mechanism, an attempt to connect to something familiar in a very unfamiliar place. A few times in the camps I earned some points with the contra radio operators by showing them that I too had mastered Morse code. Ironically, it was in those primitive camps that I got my first look at personal computers. The contras were running Tandys. They seemed to work fine under difficult conditions, so I ordered one for myself. But for the most part, during this chaotic period, electronics and ham radio remained, for me, in hibernation.

Next came two years in the Spanish Basque country. In this post, it wasn't really work that kept me away from radio—it was the social life. The Spaniards are very friendly, very social people. I found myself out with friends most evenings. I did have the Drake 2-B in operation—I'd occasionally listen to the European amateur radio chatter, but that was it. My return to amateur radio wouldn't take place until I arrived at my next post, Santo Domingo.

CHAPTER 3
TROPICAL REBIRTH
HAM RADIO IN THE DOMINICAN REPUBLIC

"I guess today's experimenters build things in software, without ever touching a soldering iron. The hocus pocus is inside the program. It's cleaner this way—nothing to burn or zap, and you don't need a voltmeter.

What happened to home-brewed and breadboarded circuitry? Where's the joy of mechanics and electricity, the creation of real things. Who are the tinkerers with a lust for electronics?"

—From "Silicon Snake Oil—Second Thoughts on the Information Highway" by Clifford Stoll, Doubleday, 1995 page 75

In June 1992, the State Department sent me to the Dominican Republic. Soon after I arrived, I met Elisa, the love of my life, the girl who would become my wife. I started to settle down. A friend of mine who is still unmarried jokes that being single is hard work, and consumes a lot of time. I guess that is one reason that many radio amateurs fall away from amateur radio during their twenties: not only are they getting started with their careers, but their spare time is often burned up by the dating game. Happily for me, that game was over as soon as I met Elisa.

There was a magazine stand in the Hotel El Embajador that for some strange reason carried *73* magazine. One afternoon, after lunch with Elisa, I picked up a copy and started reading about my old hobby. A little while later, I dragged out the HT-37 and the Drake 2-B. At first, this was simply an effort to add some unusual techno-decoration to the room that I was using as a home office. I thought the old gear would look cool off in the corner. But then I threw up a small wire antenna and started listening... Then I applied for a Dominican amateur radio license... Then I figured it would be fun to make a few Morse code contacts—sort of a retro kind of thing... Practice an old skill and all that... Then I figured I needed a slightly bigger table for the gear... Then I brought some more gear out of storage... Soon, of course, I was once again fully hooked.

What had been my home office quickly morphed into a "radio shack." This term requires some explanation. In the old days (when hams were hams) their equipment was large, dangerous and noisy. They communicated with "spark" rigs—big machines that generated RF energy by using sparks. Have you ever heard a buzz on your AM radio when a thermostat in the house was kicking in or, when someone was using a vacuum cleaner? Well, that was RF caused by sparks. The crackle that you hear on AM radios on summer evenings is RF produced by the big sparks that we call lightning. Back in the day, ham radio rigs were essentially small-scale lightning generators. This beastly gear obviously couldn't be placed inside their homes, so they usually operated out of small "shacks" constructed in the backyard. My kids and I have been reading some of the old Hardy Boy adventure books, and it was through these books that I was reminded that the term probably also has some maritime roots. When they first put radio equipment on ships, the old spark rigs were so big that they had to be housed in add-on shacks up on the deck. Hence, the term "radio shack." Over the years, most hams have been allowed to move back into their houses, but we still call the room with the radios "the shack." For a while, my kids thought that all dads had "shacks," and were surprised when they were the only kids in their school whose house had this kind of room.

My Santo Domingo Shack: Drake 2-B atop the Hallicrafters HT-37.
In front of the keyboard is a Michigan Mighty Mite.

I think Marconi had the first real "shack" and it was in that shack that he invented radio communication. In his book *Thunderstruck* Erik Larson describes how Marconi's mother helped him get his shack:

> "His mother recognized that something had changed. Marconi's tinkering had attained focus. She saw too that now he needed a formal space dedicated to his experiments, though she had only a vague sense of what it was that he hoped to achieve. She persuaded her husband to allow Marconi to turn a portion of the villa's third floor attic into a laboratory. Where once Marconi's ancestors had raised silkworms, now he wound coils of wire and fashioned Leyden jars that snapped blue with electrical energy."

My Santo Domingo Shack evolved into a thing of electronic beauty. It was cool, cool in both senses of the word. In a land of heat and oppressive humidity, I could step into that shack and immediately drop from 95F to 65F. Sometimes, it seemed like you'd get a bit of freezer-like fog when you walked in, evocative of the smoke and vapor of a mad scientist's lab.

The Dominican Republic had an excellent telephone system (privately run), but it has an electrical power system (government run) that makes wartime Baghdad's look good in comparison. Blackouts were daily occurrences, so the Embassy had equipped all of our houses with big stand-alone generators. Mine was the size of a small car, and filled a bunker-like building off the side of the house. The voltage from Santo Domingo's grid would vary quite a bit, and this would affect the performance of my equipment—early on, I spent the better part of a week trying to troubleshoot a mysterious and intermittent problem with the HT-37's relays. After lots of head scratching and hair pulling, I realized that the thing worked fine when I was using the generator. So stand-alone power became standard operating procedure at my station—this added to the atmosphere of splendid isolation in my Dominican radio shack.

Center-stage of course, was the radio gear—the HT-37 and the Drake 2-B. They sat there, magnificent in their retro-ness, seeming to defy modernity with their ancient fire-bottles. The light from the tubes and the pilot lights that illuminated the dials cast strange patterns on the walls of the shack. Later, when the internet brought us all together, fans of old radios from around the world organized a "dial light competition"—the objective was to determine which of the old radios cast the nicest light patterns in darkened radio shacks. I saw this competition as comforting proof that I was not alone in my fondness for the old gear.

There was something magical about coming into the shack late at night. Santo Domingo would be hot, humid, and always a bit noisy. Dogs would be barking, there'd be some merengue music floating in from a distant party. But walking into that shack was like entering a different world. The A/C took care of all the heat and humidity, and most of the noise. Walking into that shack also provided some escape from whatever political or diplomatic problems we were facing at the moment. I'd step into that cool air and turn on the ancient receiver that I'd bought as a boy 20 years before, and begin to listen. Signals would come pouring in from around the world. On one of my first evenings on the air from Santo Domingo, the log shows contacts with Baltimore, Guatemala City, Connecticut, Italy, Venezuela, Germany, Slovenia, Russia, Czechoslovakia, and New Hampshire.

There is a kind of seasonality to long-distance radio work. There are cyclical variations in the quality of the "mirror in the sky" provided by the ionosphere. At times, the mirror is excellent, and with very little power signals bounce happily around the globe, permitting, for example, long rambling conversations with people on the other side of the planet. Other times, the mirror is completely gone. This is largely determined by conditions on the surface of the sun—by the number of sunspots. More sunspots usually mean more ionization in the earth's atmosphere, and better long-distance communication for radio amateurs. Sunspots vary in an eleven year cycle. I was lucky—I came back to ham radio when sunspots were still plentiful.

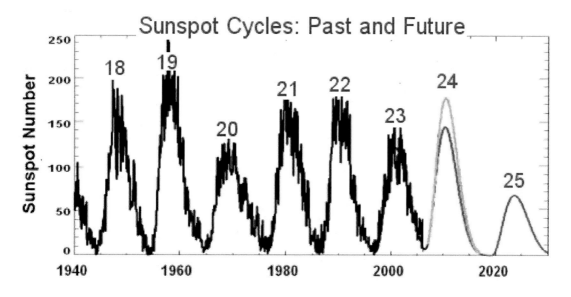

Source: http://science.nasa.gov/headlines/y2006/10may_longrange.htm

Remembering the importance of the Crystal Radio Club in my youthful experience with ham radio, I decided to seek out and join the Dominican equivalent. Amateur radio was popular in the DR—as I drove around town I'd see many ham radio antennas popping up in the urban landscape. I think geography is one reason for the popularity of the hobby—ham radio represented an opportunity for Dominicans to reach out beyond the confines of their island home. So I had no trouble finding the local hams.

Radio Club Dominicano has a very fine headquarters facility not far from where I lived in Santo Domingo. It was remarkably similar to my old New York club. There were the same stacks of musty magazines, the same weird piles of electronic detritus, the same undercurrent of geeky-ness with a hint of social awkwardness. I felt right at home.

After years in Latin America and in Spain, my Spanish was quite strong, so there was no real language barrier to impede my integration into the club. I also found that my acceptance by the members was facilitated by their memories of another American from the Embassy who had also been a member of the club. On that first night, and in many subsequent meetings, the locals spoke in almost reverential tones about "the other American," about Fred Laun. At first, I thought that Fred must have passed through recently—I was shocked to learn that he'd been there in the late 1960s. Fred had obviously made quite an impression.

Fred Laun. The club members would speak of his fanatical devotion to the hobby, of his love of marathon radio contests that would last for an entire weekend. They were still marveling at the enormous towers with directional antennas that he had boldly placed on his Embassy-provided house.

During my time in the DR, we lived through some difficult political times. There was a very problematic election, and a lot of instability in its wake. There was trouble across the border in Haiti, and several times I'd been sent out to the border region (and beyond) to keep track of what was going on. (On these trips I always brought with me some radio gear, and a copy of Carl Sagan's book "Cosmos." The book helped keep the problems I was dealing with in the proper perspective—it is always a good idea to keep the really BIG picture in mind!) But Fred had been there under much more difficult circumstances. During the late 1960s, a government with links to Castro had come to power. That caused a lot of political upheaval, and culminated in a U.S. invasion.

Gustavo Vasquez, HI8G, is one of the core members of the club. He's now a very distinguished-looking fifty-something guy with a broad smile and salt-and-pepper hair. His interests have always extended far beyond ham radio; on weekends he could be found sailing radio-controlled boats in a small artificial lake near my house. He was also into woodworking and furniture making. He's a very eloquent, very expressive person, who is not at all hesitant to tell you what he thinks. It was probably that last characteristic that got him into trouble in 1967. And in 1967, that kind of trouble could have been fatal in the DR.

At the time, Gustavo was in the University. I think it is safe to say that his opinions are a bit left-of-center now, and, as is the case with many of us, they were much further to the left during his university days. When the political firestorm hit Santo Domingo, it appears that Gustavo was on the list of people that Dominican Military Intelligence was worried about. So one day, he disappeared.

It must have been someone in the club who turned to Fred Laun. In the Embassy, Fred's job was in the public affairs area—cultural exchanges, press relations, etc. He really had nothing to do with what are called politico-military activities. But in the DR, just about anyone working in the American Embassy had influence and entrée, and was in a position to help in situations like this. To make a long story short, Fred Laun saved Gustavo. That's why they were still talking about him in reverential tones when I arrived on the scene 20 years later. They spoke of his ham radio skills and prowess, but the real reason that they were still talking about him was that he'd really become their friend. The common ground of ham radio helped bridge all the linguistic and cultural gaps, and (as would happen with me) had pulled Fred into their circle of friendship, into their cloud of solder smoke.

During one of these reminiscence sessions, Gustavo told of how, when it came time to leave, Fred had given him his beloved Vibroplex bug. (A bug is a mechanically complex Morse code key that uses weights, inertia and spring action to automatically create fast and precisely timed dits and dahs.) Gustavo mentioned that over the years some of the parts and screws had been lost, and that he wanted to restore it to operating condition. I had an old Vibroplex… Parts of mine are now mixed in with Fred's in a place of honor in Gustavo's Dominican shack.

All radio clubs have their "characters," people who have somewhat eccentric personalities. Pericles Perdomo was in this category. Pericles (pronounced Per-EEEE-clays) was in his late fifties when I met him. Beyond rotund, Pericles was almost spherical. It wasn't really obesity as much as roundness. He was like a beach ball with appendages. Like the Michelin Man or the Pillsbury doughboy, but shorter, and without their agility.

Pericles had been in the club for many years and had a sort of friendly but antagonistic relationship with most of the members. Dominican men have a tendency to kind of harass and tease each other in a friendly, mock-hostile kind of way, and Pericles was obviously a prime target for this kind of interaction. They teased him about his weight. They accused him of occasional unauthorized borrowing of the club's magazines (guilty!), and of being a bit stingy with parts in his enormous collection of radio equipment and components (he was never this way with me). Pericles didn't make it to all the club meetings, so when he did waddle in, it was a sort of a special occasion. The guys loved it when he came because, 1) they were truly glad to see their friend, and, 2) his arrival gave them the opportunity to needle him. But Pericles was smart and quick-witted and would shoot back with gusto.

Pericles lived in a large house that looked like it had been singled out for attack by enormous spider-like Martian invaders. In the DR they don't really have zoning ordinances or neighborhood associations. People are pretty much free to do as they please with their property, and Pericles used this freedom to the max. There must have been six large towers on that house, each festooned with rotate-able antennas, and each supporting multiple wire antennas that were all interspersed and entangled in ways that would send professional engineers into antenna apoplexy. Thrown in for good measure were several satellite dishes.

I was frequently at the house, because Pericles and I had a lot of interests in common. Like me, he was passionately interested in older radio equipment. And in satellite communications. And in NASA and the space program. If you looked at the two of us, you'd probably never guess that we had so much common ground. But we did, and we became very good friends.

Pericles Perdomo, HI8P, visiting my Santo Domingo shack

Pericles' long-suffering wife had been granted control of most of the residence, but in addition to his absolute authority over the roof and all associated airspace, Pericles had dominion over a group of upstairs rooms that were, collectively, his radio shack. There was a long, narrow, almost hallway-like room in which Pericles had his radios set up. He'd have to kind of waddle in sideways to get himself into position to participate in seemingly endless radio conversations with friends around the world.

Closer to the entrance, in the same long room, Pericles had his workbench. This was extremely dangerous territory, because, like most Dominican males, Pericles seemed to view safety precautions as something for old ladies or girlie-men. One afternoon I arrived at the shack to find a totally unchastened Pericles recovering from accidental contact with the 1500 volt DC power supply of an amplifier that he had been working on. It had burned a hole in his thumb, and had sent him flying across the room. (In situations like this, a beachball physique has its advantages.) An incident like this would probably have caused some U.S. hams to think about switching from radio to macramé, but in the Dominican mindset surviving an episode like this is seen as proof of being an inherently lucky person, so safety standards would be relaxed even further.

Pericles was a packrat. Nothing related to ham radio was EVER thrown away. A large walk-in closet off the shack was filled with boxes and boxes of radio parts. If you needed it, Pericles had it.

One day I was working on my old, 1950's-era Hallicrafters HT-37 transmitter. A slip of the screwdriver caused one of the custom-made wafer switches to shatter. So there I was, in the Dominican Republic, using a 40 year-old transmitter that would be considered rare in its country of origin. And I'd broken one of its few unique, custom-made parts. A visit to the local electronics shop was not going to sort this one out.

I went to the club that week and shared with members my tale of woe. "Call Pericles," was their advice. I did. "HT-37?" he said, "Yea, I think have a couple of them out in the garage. Come over and take what you need." I was astonished. That Saturday I was over at the house with a tool box. Sure enough, I had my choice of HT-37 carcasses to choose from. I extracted the needed part— Pericles' enormous junkbox got me back on the air.

I had no intention of following Pericles' lead by turning the roof of my house into a Martian cobweb, but as I got deeper and deeper into radio, I began to lust for better signals. I needed to get an antenna up higher in the air.

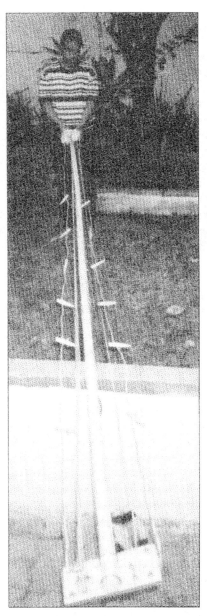

No, this isn't a new type of clothesline, it's Bill Meara, N2CQR/HI8, and his partially assembled *Dipolo Criollo*. See the complete antenna in all its glory on page 58.

I scoured the ham radio literature for designs, and the hardware stores of Santo Domingo for materials. I eventually settled on a dual-band dipole for the 20 and 15 meter bands. This would essentially be two of the dipoles that I had erected over my parents' house, but with one suspended under the other, using one common coaxial feed line. Frequent hurricanes meant that Santo Domingo has a shortage of tall trees, so my antenna would be supported at the center by a long length of PVC piping. Having never worked with PVC pipe, using it as an antenna support seemed like a good idea. Later I was to become aware of some serious shortcomings related to its, uh, flexibility.

I spent a day or so cutting the wires to the proper lengths, making a center support, connecting coaxial cable to the antenna elements and then suspending antenna number two from antenna number one. For this last task I used modified clothes pins and cable ties. This too later proved to be a less than optimum choice. But sometimes you just have to make do with what you have at hand. I had it in mind to write a little article for QST about my tropical antenna project, so I snapped a few pictures of the antenna during the construction phase of the project.

When it came time to put the antenna on the roof, I was assisted by Amado. He was one of the night watchmen assigned by the Embassy to guard my house.

Amado was about 30 years old. On my first night at the house, I'd found him hiding behind the garage door. There had been a series of attacks on night watchmen in Santo Domingo, and the poor guy was clearly too frightened to spend the night patrolling the property like he was supposed to. He was armed, but I think this just increased his anxiety—he feared (with good reason) that the bad guys might kill him so they could take his gun.

At one point in his life Amado had been a boxer, and perhaps a bit of a rogue, but that was long in the past. He was now a very devout Evangelical Christian. It was probably his wife who had changed his evil ways. When I knew him, it was hard to imagine Amado punching someone. He was a very meek, peaceful person. He really should never have been allowed to handle firearms—the security company he worked for would drop him off and give him his pistol and ammo each afternoon. Amado would then begin fumbling to get the bullets into the gun. He was sort of a Dominican equivalent of Deputy Barney Fife. After one or two accidental discharges on our crowded street, I put my foot down and ordered that all gun loading had to take place inside a workroom (with cement block walls) at the back of the yard. Hopefully that would keep the maximum body count to one.

Amado was well familiar with the roof of my house. One of the major fringe benefits of working at my place was that I allowed him to harvest the large crop of mangoes and avocados that grew on trees surrounding the house. Amado and his six year-old son Joel would come by once a week or so to harvest the crop. So when it came time to put antennas up on that roof, Amado was obviously the guy to call.

Soon he and I were up there, with the wire antenna and the long length of PVC pipe. Of course, dealing with the flexibility of the pipe became the big problem. The thing would start swaying and bowing as soon as we got it 10 feet or so in the air. It must have looked like we were struggling with some sort of weird wire and PVC octopus. This must have really provoked some head scratching among the neighbors.

At one point I was trying to explain to Amado that we had to orient the antenna so that part of the signal would go off to the northeast, towards Europe. I got a sad reminder of the kind of educational deficiencies that poor people have to live with when Amado questioned the usefulness of sending signals out to the east. "Pero no hay nada por alli William!" "But there is nothing in that direction William." Sadly, no one had ever bothered to teach poor Amado that there were other countries beyond the sea. I think I tried to gently bring him up to speed on the geography.

Eventually, with the help of some guy lines, we got the antenna in place. I must say, to me it looked magnificent up there. In my admittedly eccentric view, the two elements of that dipole looked like the wings of an electromagnetic eagle, getting set to soar out into the distance.

The new antenna resulted in an immediate improvement in both my transmitted signal and in the strength of the signals I was receiving. But the wooden clothes pins proved to be a poor choice; I noticed some changes in antenna performance whenever it rained.

I took my notes and pictures from the project and turned them into my first-ever article for QST magazine. I called it "El Dipolo Criollo" (Roughly "The Dominican Dipole"). I was really delighted to have an article in a magazine that I'd read as a kid.

When the August 1994 issue of QST finally came out, it provided us with a bit of additional amusement: One of the many nice things about the Dominican Republic is that people there are not at all hung up on skin color. It just doesn't matter as much there as it does in other places. Amado happened to be a dark-skinned Dominican. My pigmentation DNA comes via Ireland. The amusement resulted from my failure to adequately label the photos. One of them showed Amado holding the antenna just before it went up on the roof. The editors thought that he was me. The caption proclaimed: "No, this isn't a new type of clothes line. It's Bill Meara, N2CQR/HI8 and his partially assembled Dipolo Criollo." I imagine that this photo must have been a bit pleasing for those who keep an eye on the "diversity" of QST's authors. No harm done. Amado got his picture in a magazine. And now in a book.

As simple as it was, that new antenna greatly increased my ability to talk to people in distant locations. But you know how it is. You want more. You want to go farther. But the earth is round and you quickly find that you can't go any farther. If you do, you are coming up back around the other side. So, if you are a true long distance man—a true "DXer"— you find yourself thinking about going "where no man has gone before." That's right Trekkies, we're talking outer space!

I suppose at this point, some readers from outside the fraternity of solder and circuits are already stunned by the geeky-ness of all this. But hold onto your hats my non-techie friends, for we are about to enter the zone of maximal geeky-ness: The Amateur Radio Satellite Program, the home of the proverbial space cadets.

Radio amateurs had a satellite in orbit very soon after Sputnik: On December 12, 1961, OSCAR 1 (Orbiting Satellite Carrying Amateur Radio) went up in some extra space on a rocket that launched from Vandenburg Air Force Base. Over the next four decades, some 70 amateur radio satellites have been placed in orbit.

This mixing of amateur radio and space vehicles shouldn't come as a surprise. Guys who are interested in radio and electronics are also going to almost always be interested in space and astronomy-related technical challenges. And many of the guys working in astronomy or on the space

program will also be hams. This overlap of interests goes way back in ham radio. I think the best illustration is the Elser-Mathes Cup.

Sometime during the 1920s, Colonel Fred Johnson Elser, a radio amateur, met the founding father of the American Radio Relay League, Hiram Percy Maxim. Maxim was an eccentric inventor with a wide range of interests, Mars being among them. He actually owned a globe of the red planet. During their meeting, Elser and Maxim must have talked about ham radio… and about Mars.

Colonel Elser was at the time stationed in the Philippines (then under U.S. rule). Back in the islands, in 1928 he visited another military ham, Lieutenant Commander Stanley Mathes. Perhaps with Maxim's Mars globe in mind, the two of them cooked up the idea of offering a "unique trophy" (unique indeed!) for the first two-way communication with Mars. Working with the Igorot wood carvers of Baguio, Elser and Mathes had the trophy made. It has a base symbolizing the Earth. Several men hold a cup above the base. In a 1969 QST article, Elser explained that, "The bowl is Mars, and the standing men are the amateurs who bridge the gap of space." The ARRL claims that it is only a coincidence that the "men" in the trophy resemble "the large eyed hominids of alien abduction lore." Yea, sure….

There are also purely technical reasons for the ham radio-outer space overlap: While the fickle nature of our ionospheric mirror adds an important element of uncertainty and romance to long distance communication (you never really know who you are going to talk to), sometimes (especially when sunspots are scarce) you find yourself wishing for a more reliable repeater in the sky. Enter the amateur radio satellite.

As a teenager I'd tried to use my trusty Drake 2-B to listen in to the transmissions from OSCAR 6 or 7, but I had no luck. Like most of our fleet, the spacecraft that I was listening for was in low earth orbit. It was only 150 miles or so above the surface. At that height, Kepler's laws mean it has to move at about 17,000 miles an hour. It goes around the world in about 90 minutes, and, for an earth-bound observer, from horizon to horizon in about 10 minutes. So knowing precisely when to listen was very important. Today there are computer programs and web sites that tell you exactly where the satellites are, and precisely when they will pass overhead, but back in 1974 all we had were some difficult-to-follow charts from the latest issue of QST, and some sort of cardboard cut-out device called an OSCAR Locator. I always ended up listening at the wrong times, and possibly on the wrong frequencies.

But I had better luck later on: Late in 1993 I started picking up signals from space. First I heard the beacon of a satellite RS-10, an amateur device attached to a Soviet navigational satellite. I also monitored U.S. amateurs talking through an orbiting device known as AO-21. But it was all still hit-and-miss—I was still having trouble knowing exactly when the satellites would be over my location. At this point I was depending on satellite schedules that would from time to time appear in the ham radio magazines.

The solution to this problem came in the form of a very neat bit of software called Orbits II by Roy Welch, W0SL. Even on my clunky old Tandy 286 computer, ORBITS II displayed on the computer screen a Mission Control-like map of the world. After you inputted the orbital characteristics ('the Keplerian elements') of the satellite of interest, it appeared as a little blip on your new outer-space radar screen. The guess work was over—this was geek heaven. The display itself was mesmerizing—it was amazing to sit there and watch the satellites soar around the globe.

In Santo Domingo, I was not the only one interested in the satellites—Luis Ernesto, HI8LEZ, also wanted to get into space. He was a very enthusiastic young radio amateur. He is a member of one of the families of merchants in the DR who trace their ancestry to Lebanon and Palestine. When their relatives first in arrived from the Middle East, the region was still part of Turkey's Ottoman Empire, so they were labeled "Turcos" by the locals. The name stuck, and other Radio Club members often referred to Luis Ernesto by this semi-pejorative term. They called him El Turco. (Again, this kind of teasing was just part of the antagonistic game played among Dominican friends.) After I shared with Luis Ernesto the wonders of Orbits II, the Dominican Space Program took off in earnest.

Listening was fine, but for radio amateurs, the real objective is always two-way communications, so I began to work towards an actual contact via satellite. The Russian RS-12 Satellite seemed like the obvious candidate. It was designed to allow ordinary radio amateurs to engage in satellite communication without the use of any complicated or specialized equipment. You would transmit up on the amateur 15-meter band (around 21 MHz) and if the satellite received your signal, it would send it back to earth on the 10 meter band (around 28 MHz). You could use either Morse code or single sideband voice modes.

My ancient Hallicrafters HT-37 and Drake 2-B were ideal for this application. Most radio amateurs had long ago switched to combined transceivers that automatically send and receive on the same frequency, but in a happy bit of radio irony, my very obsolete '50s/60's station—with its completely separate transmitter and receiver—was ideally suited for up-linking on 15 and down-linking on 10 meters.

On January 9, 1994, I made my first contact through a satellite. The spacecraft was RS-12 and the fellow I spoke with (via Morse code) was Bob, W9JOP, in Indiana. Some amateurs say that your first contact through a satellite is as thrilling as your first on-the-air contact. I think there is an element of truth in that claim. I got a real kick out of sending my signals through that satellite.

Of course, this was the start of another form of the ham radio addiction. I wanted to make more contacts with the satellite. I wanted to use voice mode. I wanted to try other satellites. Soon I was making voice contacts through RS-12. This was a bit eerie, because as you spoke into your microphone, you would be monitoring your own signal (in your headphones) as it came down from the satellite. In normal ham radio contacts, the receiver is temporarily turned off or muted when you transmit, but with the satellites, because you were transmitting and receiving on different bands, you could send and receive at the same time, just as you do when speaking on the telephone: full duplex.

By now I was regaling the club members with details of my outer space adventures. Luis Ernesto was itching to make some satellite contacts, so we consulted ORBITS II and found that RS-12 would be over the Dominican Republic at 1:30 am local time. We agreed to stay up and attempt a Santo Domingo-to-Santo Domingo contact through the Russian satellite. We lost a lot of sleep and I was a bit of a zombie the next day at work, but Luis Ernesto made his satellite contact. That was great. But there was one other guy in the club who needed—and I mean REALLY NEEDED—a satellite contact.

One of the many things that the guys in the club teased Pericles about was his failed efforts to make a satellite contact. Because of his love for the space program, Pericles had always been very interested in the amateur satellites. Over the years, he'd spent thousands of dollars on sophisticated satellite receivers and transmitters, and on antennas that he would aim (in vain) at the heavens. It seemed like every time QST magazine would announce the launch of a new satellite, Pericles would begin the gathering of the needed gear. Some of the equipment was fairly specialized and exotic. And with each new satellite he would fail to make a single contact. He'd try and try, with no success. Eventually the satellite in question would run out of power, or fall into the ocean, and Pericles would still have not a single contact to show for his efforts. The guys at the club would tease him. I decided to help him.

I knew what was wrong. First, Pericles was trying to do something that was technically difficult for anyone, and he was trying to do it through a language barrier. Almost everything written about the satellites was written in very technical English. Pericles had some English and could make his way through most of the articles, but this was the kind of activity where just one misunderstanding would lead to failure. Second, Pericles had been trying to communicate through relatively sophisticated satellites, birds that operated on the Very and Ultra High Frequency bands and that needed precisely aimed high-gain antennas. During the years that he was trying, no one had done what the Russians had done: put up a satellite for everyman. I knew that RS-12 would be Pericles' satellite salvation.

As with Luis Ernesto, Pericles and I consulted the computer program and found a mutually agreeable pass. I went over to his house a few hours before the event, and talked him through it. The tricky part was for him to find his own signal in the satellite's downlink passband. Once he did that, I could adjust my transmit frequency so that my voice would be coming down on the same frequency as his—then we could communicate. To complicate things a bit, the frequencies would be shifting around because of the Doppler effect. So this was by no means a sure thing. We were both nervous. We didn't want another failure. I reminded Pericles that we only had a ten minute window in which to make contact. For the contact to count we'd have to exchange call signs and signal reports. Before I left his house and returned to my shack across town, I told him that if he got into trouble and didn't know what to do, he should call me on the phone, and I'd attempt some on-the-spot coaching. Fingers crossed, I headed home.

It was around 11 p.m. on January 31, 1994. As soon as I could hear the beacon from the satellite I started tuning around, listening for Pericles' familiar voice and call sign. Periodically, I'd issue a call for him. The clock was ticking and the satellite was rising up from the horizon.

As I started to call Pericles, I heard something strange in my headphones. It was an oddly familiar ringing sound, and it was definitely coming down from the satellite. I tuned it in a little better—the Doppler shift had caused some distortion. Finally, I got it tuned in and recognized what it was. A telephone was ringing. How strange! It took me a few second to realize that it was my telephone that was ringing. The headphones, combined with the whirring of the air conditioner, had prevented me from hearing the ringing directly, but I'd pushed the transmit button on my microphone, and the mic had picked up the sound of the telephone. The HT-37 sent the rings up to the Russian satellite, which sent them down to my Drake 2-B and into my headphones. I pulled off the headphones and picked up the phone. Sure enough it was Pericles. I joked that in a certain sense, he'd already made a satellite contact.

I quickly gave Pericles a couple of pointers, then it was back to the radios. Finally I heard him. I put my signal on top of his, and gave him a call. He heard me! We exchanged the necessary information. Pericles had done it! Better yet, using the few minutes still available on that pass, he went on to speak to a station in Panama. Pericles had well and truly joined the Satellite Communicator's Club. The teasing was over. He was really pleased. I got him a special certificate to mark the occasion. That was one of my most rewarding amateur radio contacts.

There was another Russian satellite up there. This one was called RS-10. It was slightly more difficult to use, because its uplink frequency was in the VHF range, around 145 MHz in the amateur 2 meter band. Like RS-12, downlink was on the 10 meter band. Thoroughly hooked on satellites, I couldn't resist the urge to add RS-10 to my orbital repertoire. Unfortunately, I had very little VHF equipment, and homebrewing gear for this frequency range was relatively difficult. But I did have that old Yaesu rig that my dad had bought me... Many years before, my dad had given me a small Japanese-made two meter transceiver, a Yaesu Memorizer. (This was the rig that I'd used to load up my rucksack while preparing for Special Forces school.) I'd never really taken that rig seriously. I had some attachment to it because my dad had given it to me, but deep down I've always been essentially a short-wave, high frequency (HF) band radio amateur. I'm most interested in the HF frequencies that permit direct, world-wide communication. For me, the VERY High Frequency (VHF) bands always seemed kind of exotic and quirky. But in the world of ham radio satellites, VHF and U(ultra)HF are where the action is. So for my RS-10 efforts I decided to press the Yaesu into service.

The RS satellites used either single sideband voice, or Morse code. The Yaesu rig was designed for Frequency Modulation (FM)—this mode was strictly forbidden on the satellite. Getting the Yaesu to transmit SSB would have probably been more difficult than starting over with a scratch-built rig. But getting a rudimentary Morse code signal out of it would be relatively easy. Other enthusiasts pointed out that simply by using the microphone's push-to-talk switch as a telegraph key, a crude Morse signal could be transmitted. I wanted to make my signal a bit less crude so I opened up the transceiver and installed some wiring that would allow me to use my telegraph key to turn on and off the final amplifier stage. Voila! 2 meter Morse. Now there were two Russian satellites cutting into my snooze time.

I never had any trouble finding people to talk to through the satellites. No matter the hour, there always seemed to be other radio fiends out there waiting for one of these satellites to pass overhead. But even if there had been a shortage of geeky interlocutors, the ingenious Russian engineers had installed a feature on these satellites that guaranteed that you'd always have someone to talk to: If all else failed, you could talk to the satellite itself.

The designers may have been thinking about the computer in Stanley Kubrick's movie *2001, A Space Odyssey* but the robots on RS-12 and RS-10 were far less eloquent than HAL. Both these spacecraft had circuitry that would respond to calls, but only to Morse code calls. You had to be on precisely the right frequency, and you had to transmit precisely the right call at the right Morse speed, but if you did, you were rewarded by hearing the spacecraft call you (using your call sign) and issue you a contact number. The robot even followed ham radio tradition by ending the contact with the traditional "73" (telegraph-ese for "Best Regards.") Through an earth-bound agent, the robot would also send you (via the mail) a QSL card confirming the contact. My old Hallicrafters HT-37 and Drake 2-B combo made contact with the RS-12 robot. The Yaesu Memorizer that my dad gave me got me the contact with the bot on RS-10. Geek chic.

There was another person who got involved in our little Dominican astronomy/space communications program: A fellow named Antonio alternated with Amado as night watchman at my house. Like Amado, Antonio was stuck in dire poverty. But Antonio didn't seem to have the kind of religious and family support that helped Amado deal with his fate. Antonio was a much more somber person than Amado, and it took me longer to get to know him.

I've been getting up early in the morning ever since my army days. Oh-five-hundred is still my standard wake up time. My return to ham radio coincided with a resurgence of interest in other long-dormant technical and scientific pursuits. Astronomy was one of them. When my friend from the Embassy, Paco Scanlan, announced in the Embassy newsletter that he was selling a 4.5 inch reflector telescope, I plunked down the requested fifty bucks. That was definitely money well spent, because that telescope soon became a window on the universe, both for me and for Antonio.

I'd get up at five. While the coffee brewed I'd set up the telescope in the front yard. At first Antonio just kind of sat there and watched. But one morning he seemed a bit intrigued so I offered to show him what I was looking at. I don't remember what it was—it may have been Jupiter, or Saturn or the Pleiades star cluster—but whatever it was it had a big impact on Antonio. "MEE-ER-DA William! Que diablo es eso?" (Roughly: "SHEEEEEIt William! What the devil is that?") Antonio quickly became my partner in astronomy. It was great to have someone out there in the yard who was so captivated by what I was doing. I tried to get Amado and others involved, but Antonio was the only one to really get interested.

Sometimes I'd take out the telescope before going to bed. Antonio would be out there and would take a look through the scope. I'd turn in, but he'd be out there all night, looking up, studying the stars. He'd still be out there in the morning when I returned. He became familiar with the constellations and the brighter stars. I could tell that he was becoming a real amateur—he was developing a real love for astronomy and for the stars. I got the feeling that this interest was filling some sort of gap in his life, perhaps letting him escape a bit from his being trapped in Dominican poverty. At the very least, it was making those long nights of guard duty easier to take. My wife has always encouraged me to go out with the 'scope and look at the stars. She says it seems to put me a in a tranquil, contemplative mood. It certainly does help keep your day-to-day problems in perspective. In "Cosmos," Carl Sagan wrote: "Our feeblest contemplations of the Cosmos stir us—there is a tingling of the spine, a catch in the voice, a faint sensation, as if a distant memory, of falling from a height. We know we are approaching the greatest of mysteries."

I guess the high point of our astronomical explorations came in July 1994, when the comet Shoemaker-Levy 9 hit the planet Jupiter. I wasn't sure if we'd be able to see any results of the impact. There was a lot of uncertainty among the scientist about what would happen. Knowing that "wishful seeing" can cause an observer to delude himself into seeing things that are not really there (like the Martian canals of Percival Lowell), I decided to use Antonio as a kind of astronomical control. For several evenings before the impact, I'd pull out the scope and—without telling him what was coming—ask Antonio to study Jupiter carefully. His eyesight was much better than mine. Through my little 'scope, Jupiter was a small disk, with clearly visible stripes caused by the cloud formations. We could see Galileo's four little moons.

A few hours after the impact, darkness fell on Santo Domingo and we finally had a chance to check (directly!) for signs of the impact. I took a look, and thought—almost for sure—that I saw a small black smudge in the upper right hand quadrant. Now, to check my observation, I turned to Antonio, who was still in the dark about the impact. (There was no Discovery channel where Antonio lived.) "Hay como una mancha oscura alli arriba, a la derecha." ("There is like a dark smudge up there on the right.") Bingo! Antonio had seen it too! But was what we were looking at really the result of the comet impact?

It was still very early days for the internet, especially in a poor country like the DR. Sometime in 1992 or '93 we'd been given e-mail capability in the Embassy, but in those days to connect from home required a long distance phone call to a Compuserve node in Miami. I know, it seems incredible, but that is what we were doing. I wrote up a message to the Compuserve Astronomy forum describing what we'd seen and asking someone to let us know if we'd seen the results of the comet's impact. When I checked back a few hours later, there was a nice note from an astronomer in the U.S. confirming that what we'd observed was a scar left by the SL9 comet.

Amateur radio and astronomy are two hobbies that seem to go together very well. There are lots of people out there afflicted with this kind of dual addiction. If you are interested in signals from remote locations, well, astronomy gets you as far as you can go. My little 4.5 inch reflector easily scoops up photons from the Andromeda galaxy—that's more than two million light years away. (That means those signals started their journey long before the development of homo sapiens. Those signals were a lot older than we are.) And there are amateur radio astronomers out there—I didn't get as far as building my own radio telescope, but I was at one point a card carrying member of SARA—the Society of Amateur Radio Astronomers. And amateurs are involved in the search for extraterrestrial intelligence.

It never really gets cold in the Dominican Republic, but in December the temperature does dip down a bit. On December 12, 1994 I was up on my roof, drinking coffee, enjoying the cool air, looking up at the stars, and confirming my neighbors' suspicions about my mental stability. I would always have to alert the guard on duty to my roof-top activities. I didn't want any .38 caliber misunderstandings. On that particular morning I saw something that led to an even tighter interweaving of my interests in radio and in space.

Here is what I enthusiastically reported to the COMPUSERVE Astroforum:

I had my first sighting of a man-made satellite this morning and would appreciate some help in identifying it. I am in Santo Domingo, Dominican Republic. The object followed a North to South track a bit West of my zenith at 0930 UTC (0530 local), 12 December 1994. Can anyone identify it? Thanks. Bill

I soon learned that at that time of the morning (and in the hour or so after dusk) satellite sightings like this are very common (so common that they are a bit of a nuisance for the astro-photographers). At these hours you, the observer, are still in the earth's shadow (a.k.a. night time), but 150-200 miles up, the satellite is in bright sunshine, with all that shiny aluminum sparkling away. I don't think we ever identified what spacecraft it was that I had seen that morning. But I thought it was pretty neat. I had seen a spacecraft zoom over my house. Orbiting spacecraft were immediately added to the list of interesting things that Antonio and I would be looking for. I started checking ORBITS II for visible passes.

My log book and COMPUSERVE messages show a burst of early morning satellite watching:

2 February 1995: I got a really nice look at the Hubble Space Telescope (HST) this morning from my perch here in Santo Domingo. I woke up early and checked the computer for any satellites that might be in the neighborhood. (I'm a ham and have been trying to communicate through the new Russian Hamsat RS-15). I noticed that HST would be visible starting at around 0935 UTC (0535 local); I scrambled up onto my roof with binoculars in hand and began to scan the southern sky. There she was, right on time! I first spotted the satellite when it was in Centaurus (with the Southern Cross glimmering off to the right) and followed it through Scorpio, below Venus and Jupiter until it vanished in the east. Great way to start the day! Yet another benefit of southern latitudes! 73 and Clear Skies, Bill (N2CQR/HI8)

5 February 1995: Computer predicted near overhead pass of HST this morning. Went out and found it. First magnitude brightness. Went right over head. Came inside and downloaded HST image of Jupiter!

The Hubble Space Telescope was a frequent visitor. It was on a near equatorial orbit, so it passed nearby us just about every ninety minutes. It was big and shiny, so whenever the skies were clear during passes just before dawn or just after dusk, we'd see the big 'scope in the sky fly over our humble front-yard observatory in Santo Domingo. It was very satisfying to stand out there in my front yard with coffee mug in hand, watch Hubble fly overhead, then step inside to download the latest images from that magnificent machine. It made it seem like that was *our* telescope.)

6 Feb 1995: Went out to look at Hubble Space Telescope at 1009 UTC. Skies very dirty, garbage dump at Duquesa burning again, but HST was shining through. Track was between Cuba and Haiti, out to the north of the DR. Not as bright as earlier sightings. Keps may be a bit old. Seems to be a minute or so behind predictions. Will look at Keps files.

The note about the burning garbage dump is a reminder that while our eyes were on the sky, our feet were firmly planted in the Third World. "Keps" refers to the Keplerian elements—the mathematical description of a satellite's orbit. These have to be updated from time to time, or else the computer's predictions will lose accuracy.

7 Feb 1995: HST in the North at 1018 UTC. Notice that the bird was very bright (1 Mag) when first sighted in the northwest, but dimmed very much when it moved to the east. Could be that I was then looking through more of the dust of Santo Domingo?

9 FEB 1995. Program showed Shuttle (STS-63) and Mir appearing briefly over horizon just before dawn. Went up to roof and caught a glimpse of one of them in the east as the sky was brightening. Timing indicates it was the Shuttle. Also saw a polar low-earth-orbit satellite going north south. All between 1015 and 1030 UTC.

13 FEB 1995. Was out looking at Mars (not much to see) and Jupiter (beautiful). Caught a nearly overhead pass of the Mir. Very bright. Moving almost north to south (northwest to south east). Amazing to think that "there are people in that shooting star."

There were people in that shooting star. Indeed. And soon we would be talking to them….

It may have been that *73* magazine that I picked up in the Santo Domingo hotel (it had a number of intriguing "homebrew" radio projects). Deep, long-suppressed memories of my desire to be a real radio amateur, to emulate Shep, Stan, Bollis, and Serge, were probably involved. And without a doubt that failed receiver project of my teenage days continued to bother me. Don't get me wrong—I really loved that HT-37 and the Drake 2-B. But every time I fired them up, a little voice from the ether seemed to whisper, "APPLIANCE OPERATOR!" I needed to melt some solder. I needed to build some radios.

"Older and wiser" are definite advantages in the homebrew radio game. As a teenager, I was prone to just charge into the first project that caught my eye. It never would have occurred to me to choose projects carefully, to pick my homebrew battles. I think that was what doomed my receiver project—I'd bit off more than I could chew. Some two decades later, I took a different approach.

This time there was a lot of preliminary research. By this I mean that I spent a lot of time reading ham radio technical books and magazines. I was once again deeply addicted to the hobby, so I would have been reading that stuff anyway, but somehow I justified the expenditure of time by classifying it as "research." I did a lot of research.

A lot of my reading took place during my lunch hour. The U.S. Embassy in Santo Domingo is co-located with the official residence of the U.S. Ambassador. The stately old house is on a small hill that rises from the back of the Embassy building. Between the residence and the Embassy there is a nice garden and the Ambassador's swimming pool. Each day at lunchtime I would head out to the garden with electronic books in hand. I had a favorite tree, and the gardeners left a chair there for me. Under that tree I searched the literature for a new construction project, and continued the quest for electromagnetic understanding that I'd begun as a teenage ham.

UNDERSTANDING: THE LOWLY CAPACITOR

One of the electronic components that I found most mysterious is, in terms of its construction, one of the most simple: the capacitor. In essence, a capacitor is just two metallic plates. There is a seemingly infinite variety: The size of the plates varies. In between the plates you'll find just about everything from a vacuum to electrolytic goo. Some capacitors are a single pair of plates, others are stacks of plates. But in essence, we're talking about two metal plates with wires running to them. Simple, right?

Well, not really. Capacitors show up in just about every stage of every circuit, and perform a variety of very important functions. A capacitor can take a varying signal that is superimposed on a DC voltage, and separate out the variation. If a varying signal has found its way into a part of the circuit that needs only a DC voltage, the capacitor can get rid of the varying signal. The books would say that the capacitor "blocks DC, but lets AC through." Capacitors help get rid of the annoying AC hum that sometimes comes out of power supplies, and capacitors are used in the output filters that prevent us from illegally transmitting harmonics.

But how do two metal plates do all of that? I was especially puzzled about how two plates with no metallic connection between them could allow a signal to pass through. I'd thought that current would only flow in a closed electrical circuit. Didn't the capacitor create a break in the circuit that would block current flow?

Capacitors had me scratching my head for a long time. Some of the books I studied limited themselves to describing what capacitors do without getting into *how* they do it. Other books went a bit further and provided some useful mechanical (actually hydraulic) analogies, but still I was left with the feeling that I didn't quite get it.

It was a musty old book from 1960 that eventually provided the enlightenment that I was looking for. I probably found *A Course in Radio Fundamentals* by George Grammer in a cardboard box under some creaking table at a hamfest. One look at the cover photo and you knew it was from another era. It wasn't just the $1.00 price tag. On the cover there is a photo of a young fellow in his teens. He was seated at an impeccable workbench. A larger Volt-Ohm meter stands at the ready, and a book is being consulted. He is working on what looks like a power supply. It looks vaguely (very vaguely!) like the supply that I built for my HW-32A. His, of course, was a lot neater. And the guy himself was a lot neater. Too neat, in fact. Of course, he was a "clean-cut" fellow (this was 1960, not '68!). His hair was perfectly combed, parted, and apparently Brylcreemed. He was wearing the kind of sweater that your mother always wanted you to wear to church. And he was wearing a bow-tie. He looked like Wally Cleaver without the smirk. Well, I guess they just couldn't have used a real ham for the picture—that would probably have scared people away from the hobby.

But anyway, Grammer's book was great on capacitors. He described an experiment with a simple device called an electroscope. Two small aluminum foil triangles are suspended from a copper wire. A pocket comb (hopefully without Brylcreem) is rubbed until it builds up a static electric charge. When it is touched to the end of the copper wire, excess electrons flow from the comb into the wire and onto the aluminum triangles. The wire and the triangles now have a

60

negative charge. Because like charges repel, the two aluminum triangles spring apart, indicating the presence of an electrical charge.

Our clean-cut hero then brings in a small metal plate. It is held in a wooden frame and is not electrically connected to anything else. A wire is attached to the metal plate, and, while watching the leaves of the electroscope, our hero touches this wire to the suspension wire of the electroscope. Instantly the foil leaves droop. And in that drooping you can see the essence of how a capacitor works.

It is really simple. Electrical charge will seek to distribute itself evenly over metal surfaces. Electrons repel. They want to get away from each other. When that metal plate is attached to the charged electroscope, the electrons crowded onto the plates will suddenly see room to move. They will spread out onto the plate until charge over the entire system is equalized. As they spread out, current will flow through the wire between the plate and the electroscope. So even though you do not have a closed circuit, when you connect that electroscope to the plate, current does (briefly) flow through the wire. Right there you have the essence of how a capacitor works, how it allows current to flow even though there is an "open circuit."

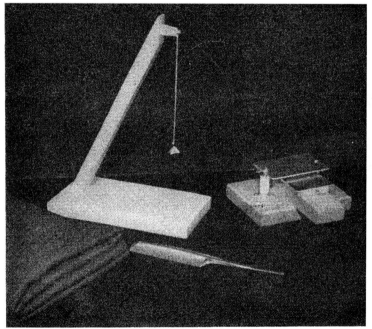

Our hero then builds a full blown capacitor. He adds a second plate that can be slid in and out from under the first plate. With the second plate out of the frame and the electroscope attached to the first plate, the system is charged—the leaves of the electroscope fly apart. The moveable plate is then connected to ground. While watching the leaves of the electroscope, the second, grounded plate is pushed into the frame, beneath the charged plate. The two plates are kept separated—there is no metal to metal contact. As the plate slides in, the leaves of the electroscope droop down, indicating that electrons have flowed out of them.

What happened? Where did those electrons go? They went into the top plate, because the addition of the second plate had increased the "capacitance." Here's how: The negative charge on the top plate had exerted an influence on electrons on the bottom plate—it had pushed electrons out of that plate, out through the wire to ground. That left positive ions in the bottom plate, these ions in turn caused the electrons on the top plate to crowd in along the bottom surface of the plate. There was now more room for electrons on the top metal plate. Electrons on the leaves saw room for expansion and flowed out through the supporting wires, into the plate.

By placing different materials between the two plates, you can have a big effect on the extent to which the plates influence each other. The material between the plates is called the dielectric.

Now, let's see what happens if the bottom plate is slid out from under the top plate. Electrons on the top plate will no longer be concentrated by the influence of the lower plate. Electrons on the plate now find themselves overcrowded and turn for more space to the relatively spacious leaves of the electroscope. Current flows out of the top plate through the wire, back into the electroscope.

Now you can see how a capacitor can separate out a DC signal from an AC signal. Replace the electroscope with a battery. For an instant after the battery is connected, current flows through the wire, but it only flows until the voltage across the capacitor is equal to the battery voltage—until the electrons in the battery-wire-capacitor system have spread themselves out. At that point, current ceases to flow.

Put a varying signal across the battery. When that signal increases in voltage in the negative direction, electrons will flow into the capacitor. When the signal starts to move toward the positive, electrons will be pulled out of the capacitor. A current meter placed between the battery and the capacitor will show that only the varying signal is causing a current—the DC voltage from the battery is not. AC has been separated from DC.

Sometimes, when you are struggling to understand something in electronics, it pays to consult a wide variety of sources. On the capacitor, I got some help from science fiction writer Isaac Asimov. In his book "Understanding Physics" he describes how an alternating current can flow through a device in a circuit that includes a capacitor (also known as a condenser):

> "In the passage back and forth from one plate to the other, the current passes through the appliance—let us say an electric light—which proceeds to glow. The filament reacts to the flow of current through itself and not to the fact that there might be another portion of the circuit somewhere else through which there is no current flow... The greater the capacitance of a condenser, the more intense the current that sloshes back and forth because a greater charge can be placed first on one plate then the other. Another way of putting this is that the greater the capacitance of a condenser, the smaller the opposition to the current flow, since there is more room for the electrons on the plate and therefore less of a pileup of the negative-negative repulsion to oppose a continued flow."

In formal electronic and physics texts you rarely find descriptions that use terms like "sloshes back and forth" or "pileup of negative-negative repulsion." But I think descriptions like Asimov's are very useful, and help you get a visual, intuitive understanding of what the formulas in the textbooks are describing.

Richard Feynman is, of course, more scientific, more mathematical, and less florid than Asimov, but in his "Lectures on Physics" Feynman sheds light on an important role of the lowly cap: "In many applications in electronic circuits, it is useful to have something which can absorb or deliver large quantities of charge without changing its potential much. A condenser or capacitor does that."

Here Feynman is talking about what we call a "bypass" capacitor. A wire carrying DC voltage to a stage of a radio receiver or transmitter may pick up some AC signals. They could come from nearby power lines, or from oscillators and amplifiers in the device itself. We don't want that AC signal on that wire. It could cause an unpleasant hum, or it could cause that stage to start oscillating on its own (not good). So we put a bypass capacitor from that wire to ground. After a brief instant of initial charge, the DC voltage on the line will see the capacitor as an open circuit, and the DC current flowing into that capacitor will halt. But when the unwanted variation from the picked-up AC signal comes along, they will flow easily into and out of that capacitor. They will see that capacitor not as an open circuit, but as a short circuit. They will flow into and out of that capacitor, and not into and out of the stage that that line is feeding. To use Feynman's word, the capacitor "absorbs" the troublesome, unwanted charge. Another nice mechanical metaphor for the bypass capacitor is the shock absorber in a car.

Capacitance is measured in Farads (a nod to Michael). A capacitor of one Farad allows one ampere of current to flow when the voltage across it is changing by one volt per second. A capacitor of one Farad will store one coulomb of charge when one volt is applied across it.

****

In addition to all my "research," I also gathered my tools. Somehow, miraculously, after all the years and all the moves to all those countries, I still had with me many of the same tools that I'd used as a kid. I still had the big old Weller soldering gun (no self respecting American 15 year-old would use a puny soldering iron when a soldering GUN was available!). I also still had the voltmeter with which I'd diced with death ("800 volts DC! Perfect!") during the HW-32A power supply project. I still had my favorite needle nose pliers.

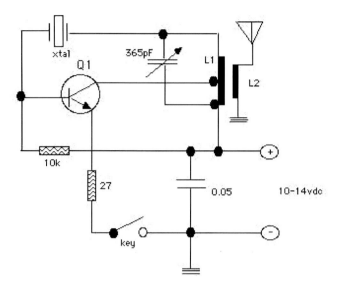

I decided to start off slow, with small projects that seemed likely to succeed. The secretary in our office in the Embassy, Mady Bullen, had an interest in ham radio that had been sparked by service in far-off places where short-wave was the only way to talk to home. She would pass me old issues of CQ magazine. It was in the March 1992 issue that I found the Michigan Mighty Mite.

It was originated by Ed Knoll, W3FQJ and developed by Tom Jurgens, KY8I. It is about as simple as you can get in a radio transmitter: just one stage, a crystal controlled oscillator.

An oscillator is basically an amplifier in which some of the output signal is fed back into the input. If you provide enough feedback in the right way, the amplifier will "take off" and begin generating a signal. The howl you hear when the microphone of public address system gets too close to the speaker is this kind of signal. The speaker (the output) is sending energy back to the input (the microphone) and what was an amplifier turns (annoyingly) into an oscillator. In this case it is an audio frequency oscillator because all the filters and tuned circuits in the PA system are built for the audible frequencies. But the same thing will happen at radio frequencies. That's what the Michigan Mighty Mite is all about.

I put the thing together using parts obtained from the Santo Domingo Radio Shack store. The resonant circuit used a coil that was just some wire wound around a discarded plastic 35mm film container. Homebrew radio projects rarely work the first time you power them up. I had to fidget with this thing quite a bit—obviously there wasn't enough feedback. I had my Drake 2-B on and tuned to the crystal's frequency. As I poked around on the little circuit board, I suddenly heard a little chirp from the 2-B. There it was! The little device that I had put together was producing radio frequency energy on the 40 meter band. Hooray! The joy of oscillation! Now I felt like I was truly in league with Faraday and Marconi, with Shep, Stan and Bollis, and with Serge! Hilmar would have been proud of me (but he still would have been horrified by my sloppy wiring).

I never was able to talk to anyone with that little device—the power output was very low, and my antenna for the 40 meter band was very poor. But it didn't really matter. I had had my first real success at homebrewing a piece of ham radio gear.

UNDERSTANDING: RESONANCE AND OSCILLATION

Even though that Michigan Mighty Mite was very simple, there were some fundamental aspects of it that I did not understand. What did that coil wound on the plastic 35mm film container and the capacitor connected across it actually do? What did that crystal do? How did that little combination of parts produce a radio signal?

Some of the books just kind of breezed through the role of resonant circuits in oscillators. They would point out that sometimes the coil and the capacitors determined the frequency of oscillation. In other cases these circuits—through something called the flywheel effect—are used to clean up the output of amplifiers. But I wanted to know *how* they did all this.

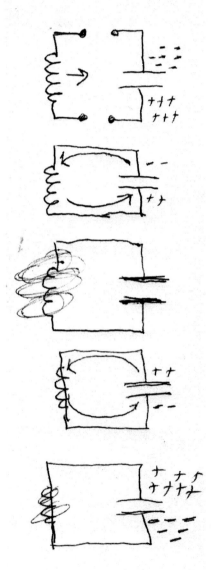

For me, a glimmer of understanding came via War Department Technical Manual TM 11-455 "Radio Fundamentals" dated July 17, 1941. I guess this was another hamfest purchase. The chapter "Vacuum Tube Oscillators" begins with a discussion of mechanical oscillation. Imagine an ordinary spring lying on a table. Hold one end steady. Compress the spring and let it go. It springs out to its uncompressed length. Now attach a heavy weight to the free end. Compress the spring again and let it go. This time the spring will be slower in expanding, and when it gets to its uncompressed length the inertia of the weight will cause it to expand beyond this point. The weight will bob back and forth. It will oscillate.

With this mechanical model in mind, let's switch to electrical resonant circuits. Here we have a parallel LC circuit. Think of the capacitor as the spring and the coil (the inductor) as the weight that provides the inertia. Start out with a charge on the capacitor (imagine a compressed spring). When the coil is connected across it, the capacitor will start to discharge through the coil (the spring is released). This will set up a magnetic field around the coil that will—at first—resist the current coming from the capacitor (think of the inertia of the weight resisting the initial expansion of the spring). Eventually the capacitor will be almost fully discharged, and the current from it will begin to decrease. But now the magnetic field around the coil will begin to change in such a way as to keep the current flowing. The coil will now start charging the capacitor with the opposite polarity (imagine the weight swinging out beyond the point at which the spring is fully extended).

Energy is being handed back and forth from the electrical field in the capacitor to the magnetic field in the coil, just as energy was being handed back in forth from the kinetic energy in the weight to the potential energy of the compressed coil. Just as the frequency at which the weight bounced at the end of the spring varies with the characteristics of the weight and the spring, the resonant frequency of the LC circuit depends on the electrical values (capacitance and inductance) of the coil and the capacitor.

Another very helpful mechanical analogy is the playground swing set. Think of the inertia of the moving child as the inductance (it requires energy to get going, and once moving tends to keep moving). Think of gravity as the equivalent of the capacitance. The coil in the LC circuit will dump energy into the capacitor until the capacitor puts a stop to the current flow—gravity and inertia interact the same way in the playground swing set. Like an LC circuit, that kid and his swing will have a natural resonant frequency. It will be based on his or her weight, and the length of the swing's chains. As with the LC circuit, if the input signal (in this case your pushes) matches the resonant frequency of the kid/swing combination, the swinging will be maximized, just as the circulating current is maximized in an LC circuit when the input signal is at the circuit's resonant frequency.

In the Army's "Principles Underlying Radio Communications" (1921) the authors use a military example: soldiers marching on a bridge:

> *"Many mechanical examples of resonance might be cited. It is a well-known fact that the order to 'break step' is often given to a company of soldiers about to pass over a bridge. Neglect of this precaution has sometimes resulted in such violent vibrations of the bridge as to endanger it. This is especially the case with certain short suspension bridges. When a shock is given to the bridge it vibrates, and the frequency of the vibrations—that is, the number of vibrations per second—is always the same for the same bridge, whatever the source of the shock. The frequency of vibration is analogous to the resonance frequency of the circuit. For if an impulse is applied to the bridge at regular intervals, tuned so that the number of impulses per second is exactly equal to the number of vibrations natural to the bridge in the same time, violent vibrations may be set up, although the individual pulses may be small. In fact, when the bridge is vibrating the impulses need to have only just force enough to overcome the frictional forces, and thus keep the vibrations from dying away. The much greater force involved in the vibrations themselves correspond to the large voltages acting on the coil and condenser [capacitor]. The voltage on the condenser is of the same nature as the large forces which exist in the beams of the bridge when they are stretched, while the voltage on the coil corresponds to the considerable momentum of the moving bridge. The small force of the impulses given the bridge correspond to the small applied EMF in the electrical case."*

In London I lived very close to the Albert Bridge, which crosses the Thames. There are signs on that bridge ordering marching soldiers to "break step" when going across. (When I was in the army, the command was "Route Step, March!") The Discovery Channel's "MythBusters" performed an experiment in which they tried to destroy a bridge using the principle of resonance and marching soldiers. Apparently it is hard to do in practice (good thing!).

Stability is very important in resonant circuits. These combinations of coils and capacitors determine the frequency at which your rig will be transmitting. Obviously you don't want that frequency to be shifting around in the course of an on-the-air conversation. Unfortunately, stability can be hard to come by when using coils and capacitors. Even slight changes in temperature will change the physical dimensions of the components, and this will change the resonant frequency of your circuit (and your rig).

Mother Nature provided radio amateurs with a very simple way to obtain frequency stability: quartz crystals. Quartz has "piezoelectric" properties: If you squeeze it, a slight electric charge appears across the rock. And (very importantly) if you put a slight electrical charge across it, the crystal will physically expand and compress with changes in that charge. It was discovered that quartz crystals could be cut to resonate at very specific frequencies. This became critically important in World War II when the military needed "rock steady" frequency stability for networks that involved stations operating in wildly changing temperature extremes. Drifty, unstable oscillators built with capacitors and coils just wouldn't work. Later, crystals became useful in the effort to prevent radio amateurs from drifting out of their authorized frequency bands. Until 1972, holders of the American Novice Class license were limited to the use of crystal-controlled transmitters. The crystals used in WWII (and by radio amateurs to this day) were usually plug-in devices. A sliver of quartz ground to a specific frequency was placed between two metal plates wired to connecting pins.

For me, the crystals were easier to understand than the LC circuits. They are similar to tuning forks: You hit them with an electrical charge that causes them to physically contract. Then, like a tuning fork, when they try to bounce back to original form, they overshoot a bit. And so on, all the while producing an alternating electrical signal at a frequency determined by the physical parameters of the crystal.

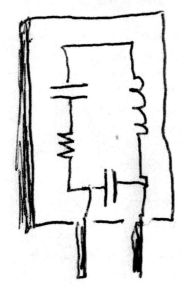

"Electric Radio" is a wonderful little magazine devoted to old tube-type ham radio gear. All the articles are by fellow enthusiasts. These kinds of publications often provide the most understandable explanations of electronic circuits and mysteries. An article by Bob Dennison, W2HBE, in the December 1994 edition allowed me to relate the physical, mechanical properties of the crystal to its electrical action.

Dennison wrote that we could think of the crystal as a series circuit with inductance (L), capacitance (C), and resistance (R). He said that the inductance can be seen as the equivalent of the crystal's physical mass. Think here of the inertia of the crystal, of the weight of the tuning fork, or the weight of the kid on the swing. The capacitance can be seen as the elasticity or the "springy-ness" of the device. The resistance represents the internal friction of the crystal. This is what prevents our little rock from becoming a perpetual motion machine: like a tuning fork, without occasional injections of energy, the internal friction in the device will cause it to stop vibrating. Dennison's diagram shows a second capacitor between the pins. This represents the capacitance caused by the two metal plates on either side of the rock itself.

****

Even though it hadn't produced any contacts, that Michigan Mighty Mite put me on a new path in radio, on the path toward the "less is more" world of QRP.

The old Morse signal for low power or "lower your power" is "QRP," and those three letters have been adopted by fans of simple, low power radio gear. The QRPers constitute a kind of radio counter-culture. While the masses migrate to high power commercial radio gear that is ever more expensive, sophisticated, and beyond the comprehension or repair capabilities of mere mortals, the QRPers move in the opposite direction, toward simple, home-made gear that they design and build themselves. The British have been leaders in this movement. They have a very fine club called G-QRP. It was through them that I obtained a kit for a receiver.

I'm sure that memories of my failed teenage receiver project was one of the reasons I bought that receiver kit. Like the receiver that I'd attempted to build years before, this receiver was also direct conversion.

It arrived from England in good order and I quickly put it together. It worked... But there was something wrong. It had gone together TOO easily. It was more a matter of "putting it together" than of "building it." And there was something about the circuit itself that I didn't quite like. It was essentially just two integrated circuit (IC) chips, one that amplified the signal from the antenna AND converted it to audio, and another that amplified the audio signal to a level suitable for earphones. When the thing was done, I really didn't have much of a sense of accomplishment. I looked at it and thought that most of the circuitry was actually inside the two little integrated circuit chips, in those two little black boxes. I thought to myself that I had to a certain extent simply substituted small black boxes (the ICs) for larger black boxes (the modern commercial rigs). I didn't like this. Without realizing it, I was on my way to becoming a RADICAL FUNDAMENTALIST homebrewer. Soon I would find myself admiring the work of early radio amateurs who had fabricated their own vacuum tubes, and feeling slightly guilty about using "store bought" transistors.

I wanted to build radios "from scratch" and I wanted to use "discrete" components, individual parts, not the pre-packaged circuits that constitute modern integrated circuit chips. Kits and chips were fine for some people, but for me, in my radical fundamentalist mindset, they represented a kind of halfway house, a lukewarm middle ground between the ignominy of "appliance operator" status and membership in the hardcore homebrewer elite.

Ironically, Santo Domingo turned out to be a very good place for homebrewing radio gear. It is very common to hear radio amateurs bemoaning the difficulty of acquiring parts for their radio projects. Sometimes it seems that if the 7-11 shop on the corner doesn't stock T-50-2 toroidal inductors, some guys just throw up their hands claiming that "it's too hard to get the parts!"

You'd think that in a poor country like the Dominican Republic it would be even more difficult to acquire components, but—and this is the ironic bit—this is not true. In poor countries, people still repair things. In the rich countries, we just throw out malfunctioning TVs and stereos. Not so in the Third World. There, things still get repaired. So there are businesses devoted to supplying parts to the repairmen.

In Santo Domingo, these businesses were all near the street "16 de Marzo." It is a fairly chaotic place in the older part of town. On a Saturday afternoon it would be teeming with people, all of them engaged in some form of electronic commerce. There were several large electronic retail stores—they seemed to specialize in the very large audio amplifiers that are needed to ensure that the DR's Meringue music vibrates not only eardrums, but entire city blocks. There were also shops that repaired devices damaged by the country's very rickety electrical power system.

Most important for my purposes were the parts suppliers. These were usually tiny little stores. Out front there were colorful, hand-painted signs saying RESISTORES! CAPACITORES! I'd squeeze my way in and find a setup that looked almost identical to the thousands of small, neighborhood food shops of Santo Domingo. It was disorienting at first, because it seemed impossible that electronics components would be sold here. "Digame Señor" the young girl behind the counter would say. ("Tell me what you want Sir"). I'd quietly ask for the six 2.2 kilo-ohm resistors that I needed. I fully expected a baffled look from the girl, but without batting an eye, as if I'd just asked for nothing more exotic than a pack of Marlboros, she'd ask, "Kilo o mega?" (Kilo-ohms or mega-ohms?") "Kilo por favor." "Cuantos vatios?" ("How many watts?"). In that last question she was asking me about the power handling requirements of the resistors that I was looking for. "Un quarto de vatio, por favor" ("A quarter watt please"). She'd go look for the parts, put them in a little plastic bag, I'd pay up and I'd be off to the inductor merchant next door.

The electronic trade had spilled out onto the street. Out there on the narrow sidewalks were some people offering some very specialized services. At one point I managed to burn out the transformer on one of my transmitters. The transformer is a big, ten pound hunk of iron and wire. Inside there are hundreds of turns of magnetic wire. This is the device that by way of Einstein's Special Relativity turns the 115 volts from the wall socket into the voltages needed by the various circuits in the transmitter. Mine had burned up. Getting a replacement in the States would have been a complicated, big-bucks operation. But not in Santo Domingo. It must have been the guys in the radio club who alerted me to the existence of the "transformer specialists" who operated on the sidewalks of 16 de

Marzo. The owner of the resistor shop pointed one of them out to me. The street corner setting made the whole thing feel sort of illicit, more like a dope-deal than a transformer repair. I half expected the repairman to be whispering to passers-by, "Pssst... Hey buddy... Transformers rewound! RF chokes repaired..." But of course there was nothing illicit about it. The fellow was providing a needed service. He did very good work. I don't think he knew about the physics of Faraday and Einstein, or of Maxwell's equations, but he knew how to rewind the transformers that made use of their discoveries.

UNDERSTANDING: SERIES AND PARALLEL TUNED CIRCUITS

"Very much later, when I was doing experiments in the laboratory--I mean my own home laboratory, fiddling around--no, excuse me, I didn't do experiments, I never did; I just fiddled around. Gradually, through books and manuals, I began to discover there were formulas applicable to electricity in relating the current and resistance, and so on. One day, looking at the formulas in some book or other, I discovered a formula for the frequency of a resonant circuit, which was $f = (1/2)\pi\sqrt{LC}$, where L is the inductance and C the capacitance of the... circle? You laugh, but I was very serious then. π was a thing with circles, and here is π coming out of an electric circuit. Where was the circle? Do those of you who laughed know how that comes about?" —- **Richard Feynman, April 1966 lecture to the National Science Teachers' Association. Reprinted in "The Pleasure of Finding Things Out" By Richard P. Feynman, Perseus Publishing, 1999**

In Santo Domingo I started using notebooks as part of my effort to understand electronics. I soon realized that these notebooks were almost as important to my effort as the text books, handbooks, and technical manuals that were accumulating on my bookshelves.

I use black and white "Marble" composition notebooks, the kind with the multiplication tables inside the back cover (mine now have rectangular holes in the back covers, because these tables were cut out and used in Billy's fourth grade efforts to master multiplication). As I'm trying to understand something, I'll fill up the pages with schematics, formulas, math scribbles, graphs and verbal explanations on the topic I'm trying to grasp. Looking back at the Santo Domingo notebooks, I see many instances where a bold "Eureka!" was scrawled across the top of the page, only to be scratched out and replaced with a disappointed "No" when I found out that my understanding was, for one reason or another, incorrect.

I strongly recommend the use of notebooks, but I must also advise some caution: Don't let them fall into the hands of people who do not share our enthusiasm for electronics. Be especially cautious about letting co-workers see them, especially if your job (like mine) is not related to electronics or engineering, and requires a security clearance. You must realize that these notebooks will, to the uninformed, look like something produced in a cabin in Montana by the Unabomber. Better to keep the notebooks private, OK?

Looking back at my Santo Domingo notebooks, I see that I was struggling with resonant circuits—combinations of inductance and capacitance designed to resonate at one particular frequency. This was an area of inquiry that was particularly dangerous from the "what would my co-workers think?" perspective, because it involved lots of diagrams with the spring-like schematic symbol for coils, lots of sweeping arrows showing how electrons surge back and forth from coil to capacitor, lots of stuff to make an inquiring police officer suspect that he's just come across a budding Ted Kazinski.

But in spite of the risk of ridicule and arrest, I pressed on. I wanted to know why and how the combination of a coil and a capacitor could serve as a frequency selective filter. The LC circuits (L is the symbol for inductance, C for Capacitance) show up all over the schematics of radio gear, and I knew that unless I really understood how they worked, I wouldn't really be one of the electromagnetically anointed.

To really understand this, I think you have to start out by looking at what happens to voltage and current when alternating signals are placed across capacitors and inductors.

First, consider the inductor: In a resistor, as soon as you place an AC voltage across it, AC current flows through it. This current is completely in step (in phase) with the voltage. But this is not the case in an inductor. As soon as you apply the AC voltage, a magnetic field is set up around it. In our discussion of Faraday and Maxwell, we saw that a changing magnetic field creates a changing electric field. We usually think of this new field as affecting other wires (like the secondary on a transformer, or a distant antenna), but it is important to realize that this new electric field will also affect the inductor that created it. In this inductor, the fields will create "back EMF (electromotive force)" that opposes the voltage from the signal. The more quickly the input signal is changing, the greater the opposition. If you look at the waveform of a typical sine wave, you will see that the most rapid changes in voltage take place when the voltage is close to zero (rising or falling). At the peaks in the curve, the rate of change is much lower (look at the instantaneous slope of the curve). So, when the signal voltage is beginning its climb up from zero, there is a lot of back EMF and very little current flows. When the signal voltage is approaching its peak value, the back EMF begins to disappear, and current begin to flow through the inductor. We say that in an inductor, current lags the applied voltage by 90 degrees.

Now consider the capacitor: At the moment that the AC signal begins to climb up from zero, the capacitor experiences a big change in voltage. Electrons will surge onto the capacitor plates, just as they did in the experiment with the electroscope. During the portions of the cycle of the input signal in which the input voltage is changing rapidly, current will flow into the capacitor so easily that it will appear to be a short circuit. And remember: if you put a voltmeter across a short circuit, it will show zero voltage. So in the capacitor you have the somewhat counterintuitive situation in which there is current into the capacitor, but little voltage *across* it. A bit later on in the cycle, as the input signal approaches its peak, the rate of change (picture the instantaneous slope of the curve) decreases. Now the input signal starts to look more and more like a DC voltage. And remember, to DC, the capacitor looks like an open *circuit.* So at this point in the cycle current into the capacitor decreases while the voltage across it reaches maximum. In a capacitor, voltage lags current by 90 degrees.

L is the symbol for inductance, C for capacitance, I is for current, E is voltage. You can remember the lag and lead patterns by thinking of "ELI the ICE man." Get it?

Let's put a capacitor and an inductor together. How they work in combination depends on whether they are in series or parallel. Let's look at the parallel situation first: We'll use capacitance and inductance values that will yield the same reactances at the frequency we are using. Suppose for example that we are operating at 14.2 MHz and we want 100 ohms of capacitive reactance (Xc) and 100 ohms of inductive reactance (Xl). Using the formulas $Xl=2\pi fL$ and $Xc=1/(2\pi fC)$ we get values of 1.20 uH and 120.44 pF. In parallel circuits, the voltage across each element will be the same. And because we have chosen values of capacitance and inductance that will yield identical reactance values, the current through each of the components will be of equal levels. But look at the phase of the current in each element: In the capacitor, current leads the voltage by 90 degrees. In the inductor, current lags voltage by 90 degrees. So at 14.2 MHz, the current in the inductor will be of equal strength to the current in the capacitor, but the two currents will be 180 degrees out of phase. The same number of electrons flowing down into the capacitor will be flowing up out of the inductor. This is the same situation that was presented in mechanical terms in that 1941 Army Technical Manual I described earlier. There will be large currents circulating inside the tuned circuit, but net current through the parallel combination will be close to (but not quite) zero. In the old days this kind of parallel tuned circuit was called a "tank circuit" because it serves as a kind of tank to hold current. In this case 14.2 MHz is the resonant frequency. It is a bit counter-intuitive, but in this parallel tuned circuit, at the resonant frequency the impedance of the circuit will be high.

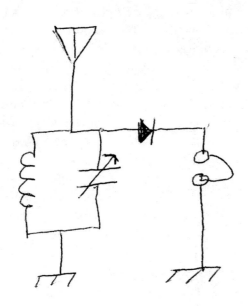

This high impedance at the resonant frequency can be very useful. Perhaps the best illustration of this comes in the form of the crystal radio that gave my old New York radio club its name. Signals of many frequencies will be coming to the receiver from the antenna. Those that are above the resonant frequency will see that capacitor as a low impedance path to ground. Those that are below the resonant frequency will see the coil as a low impedance path to ground. So above and below the resonant frequency the parallel LC circuit in effect shunts the signals away from the crystal detector and into the ground. Only at the resonant frequency will the signal see the LC circuit as a high impedance—only at this special frequency will the incoming signal appear as a detectable voltage and move through the crystal detector and on to the headphones. That's how a parallel LC circuit serves as a filter.

The inductor capacitor combination works very differently when they are placed in series, and for some reason I always found the functioning of the series circuit a bit harder to visualize.

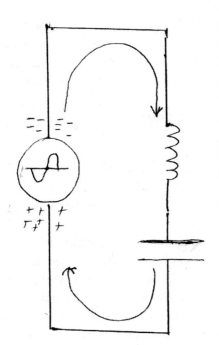

Here's one way of looking at it. Consider what happens at one part of the cycle of the input signal: As the voltage from the signal source begins to move from zero into the negative direction, the coil generates "back emf" and resists the change in current. At this point in the cycle there is a large voltage across the coil, negative at the top in our diagram. But look what is happening down at the capacitor end. At the same moment, the generator has suddenly put a positive charge on the bottom plate of the capacitor. Electrons flow out of that plate, putting a positive charge on that bottom plate. If the inductive and capacitive reactances are the same, the current flowing out of the capacitor will exactly counteract the current being blocked by the back EMF of the inductor. Because the reactive effects of coils and capacitors are always 180 degrees out of phase, this kind of counteraction will take place throughout the cycle. And if the two reactances are identical, they will balance each other out exactly and in effect the inductor and the capacitor will disappear from the circuit. But remember, this will happen only at one particular frequency. That gives us the tremendously useful filtering function of the LC circuit.

Here's yet another way of seeing it: In any series configuration, the current through the two devices has to be the same. (Picture a series of garden hoses of different diameters all spliced together. The diameter of the narrowest hose will determine the water flow through all of them, and the water flow will be the same through all of them.) But in series circuits, the voltage across each element can differ. (That is the basis for the voltage divider circuit that appears so often in power supplies and transistor biasing circuits.) And because voltage leads the current in an inductor while lagging the current in a capacitor, when the two reactances are the same (that happens at the resonant frequency) at any given moment in the cycle, the voltage across the capacitor will be equal in intensity, but opposite in polarity to the voltage across the inductor. The two voltages will cancel out. If there is no voltage across something, well, that is the definition of a short (zero resistance) circuit, right?

In both the series and parallel circuits, it is helpful to think about what happens when you move away from the resonant frequency. In the parallel circuit, the strength of the current through the capacitor will no longer be equal to the strength of the current through the inductor. So there will be some current through the parallel combination. Impedance will drop as you move above or below the resonant frequency. Depending on which way you move the frequency, the combination will appear to be either capacitive or inductive.

In the series circuit, the voltages across the capacitor and the inductor will no longer be matched. Suppose for example that we are now at a frequency at which the inductive reactance is a bit larger than the capacitive reactance. In this case, at the start of the cycle, the back EMF produced by the inductor will be less than fully compensated for by the current being pulled out of the capacitor. The combination of the two components will look like an inductor.

****

Older and wiser than I was as a teenager, I'd learned to pick my battles carefully, and to not bite off more than I could chew. So I took my time in selecting my next project, the project that would, I hoped, allow me to enter the ranks of the anointed ones, the true homebrewers.

There is a bit of ham radio conventional wisdom that says that transmitters are easier to build than receivers. I think this is debatable, but memories of my failed teenage receiver project left me inclined to accept it. So, with my Michigan Mighty Mite experience under my belt, I decided to build another transmitter, hopefully one that would actually produce on-the-air contacts.

In the ARRL anthology *QRP Classics* I found my circuit: "A VXO controlled CW transmitter for 3.5 to 21 MHz." VXO stands for variable crystal oscillator—some extra components are attached to the fixed-frequency crystal in the oscillator circuit so that you can shift the frequency around a bit. CW stands for Continuous Wave—this is the seemingly paradoxical technical name for Morse code. Why would they call "continuous" something that essentially requires the radio wave to be turned on and off? The answer lies in radio history. The spark transmitters used in the early days transmitted waveforms that were far from steady and even, far from "continuous." Then crystal oscillators came along and our radio waves became smooth and steady. So even though we were turning the transmitters on and off to transmit Morse Code, when the signals were going out (when the key was down) the waves were now "continuous."

This was a very simple transmitter. It used only five transistors, and promised to send out approximately 6 watts of power (more than enough for worldwide communication and actually one watt beyond the usual definition of QRP). The circuit was very clean and straightforward: The oscillator sent radio frequency energy to a series of amplifiers. The signal was then passed to the antenna and on to the universe. It employed only "discrete" transistors—no mysterious "black box" integrated circuits to offend my radical fundamentalist sensitivities. All the circuitry was on one simple printed circuit (PC) board. There was a vendor who would sell you a ready-made board for the rig—you'd just have to stuff the parts into the appropriate holes—but of course I disliked this laborsaving approach. From the standpoint of a homebrewing purist, I decided to go with a PC (Politically Correct) PC board: one made by hand, by me. The transmitter was designed for use on one single amateur radio band. I selected the 20 meter (14 MHz) band because of its potential for long-distance communication.

Parts gathering commenced. Most of the components came via orders to U.S. mail order vendors, but as the project progressed I did occasionally have to go down to the 16 de Marzo district. I was pleased to include in the rig parts acquired from these shops—they seemed to add a Dominican touch to the project.

The first order of business was the PC board. Here I started with simple copper-clad circuit board. The idea is that on those portions of the board where you want the copper to remain (the copper will serve in lieu of the bits of wire that were used to connect components in the old days) you cover the board up with something (tape, nail polish, special etch-resistant ink) and then put the board in an acid bath that will etch away the exposed copper. Of course, having to work with caustic chemicals added to the allure of the project. It also allows the builder to acquire one of the identifying marks of a true homebrewer: green etchent stains on the hands and clothing. (Another identifying feature: burn marks on the hands in the shape of soldering iron tips.)

I was a bit of a perfectionist on this first attempt, and I didn't like the way my first board came out. So I started over. (I later realized that my first board, while not pretty, was perfectly useable—this would later serve as the base for a 30 meter version of this transmitter.)

Holes were drilled and parts were carefully placed in their intended positions. Over the course of a few weeks, the circuit was completed. Now came the time for the feared smoke test. I powered up the rig. No smoke. Test passed. I listened for the oscillator. It was there, clean and strong. I hit the telegraph key, and the meter showed power going to the antenna. With the circuit board still sitting—sans cabinet—on the table in front of the Drake 2-B, I attempted my first contact. It was September 24, 1993. SP9JKW was calling CQ on 20 meters. As he called, I carefully made sure my transmitter was on his frequency. At the appropriate moment, I called him. Bingo! Miro in Keilce, Poland gave me a very respectable signal report. On my first attempt with the new homebrewed rig, I had spanned the mighty Atlantic.

I put the circuit board in a proper cabinet, with the control knobs arranged to my liking. I added a small relay circuit that made it easier to switch from transmit to receive (the relay would switch the antenna from the Drake 2-B and silence the receiver while I was transmitting). My purist sensibilities were bothered by the fact that I was using a store-bought power supply for the rig, so I gathered parts and built my own. Now I could say with complete honesty "Transmitter here IS HOMEBREW." You could feel the admiration and envy that this provoked, even from 3000 miles.

The radio gods must have approved of my project, because they waited until the homebrew rig was finished and on the air before smiting my beloved HT-37. At this point I was deeply hooked on radio, and I probably would have gone into withdrawal if I'd been forced off the air. As it was, when the HT-37's RF choke coil burned out, my new homebrew 20 meter rig was sitting there ready to fill the gap while the guys on the corner down on 16 de Marzo re-wound the choke.

A look at my logbook for this period provides reminders of the joy and fun of being on the air from an exotic location, using an exotic home-made transmitter and a legendary receiver. Many of the contacts were with guys that I'd talked to before—a small community of "regulars" had cropped up for me around the frequencies of my crystals. Tom James, W1HET, in Massachusetts, was one of the regulars. Following the old ham radio ethos of "my junk box is your junk box," Tom would periodically send me "homebrew care packages" containing hard-to-find radio components. Then there was Galo, HC1GC, in Quito Ecuador. Galo was definitely part of the brotherhood of solder smoke, the kind of guy you saw in front of impressive homebrew contraptions in the pages of *QST*'s "How's DX?" column. His rig was completely homebrew, and made—in the finest of ham radio traditions—of parts scrounged from discarded television sets. Galo and I soon began to complement our radio contacts with an exchange of letters and packages. We'd swap schematic diagrams and needed parts. Here in my shack in Rome I still have on display a nice Ecuadorian tapestry that Galo sent along in one of those packages.

My electronic adventures in Santo Domingo were not limited to the Morse code end of the technology spectrum. Exciting things were happening in the computer world during the 1990s and many of these developments were spilling over into the DR and into the world of amateur radio.

Dominicans love new communications technology. This intensely social group of people embraces any new opportunity to reach out beyond the confines of their island home. That is probably why ham radio had always been unusually popular in the country. That's why cell phones became popular in the DR faster than in the US. And that's why by 1995 the internet had taken the Dominican Republic by storm. No longer was I forced to use my clunky old Mosaic browser to connect to Compuserve via a long-distance phone call to Miami. The local phone company had become a fairly efficient ISP, and the DR was on-line, big time.

It didn't take long for amateur radio to make its way into those new fiber optic lines. We have a tendency to think of Voice-over-Internet (VoIP) as a relatively new innovation, but programs that allowed us to talk over the net actually arrived very early. In 1995 amateurs were experimenting with a program called Internet Phone by an Israeli company called Vocal-Tech.

VoIP digitizes the human voice and sends it out on the internet in packets. The packets are reassembled at the receiving station. While the technology is far different than that normally used on the short-wave bands by radio amateurs, when hams get on the VoIP systems, the conversations are remarkably similar to the traditional ham radio contacts.

Internet Phone was a lot of fun, but it had its problems. It seemed that there were two groups of people interested in using it: fans of ham radio, and fans of pornography. Fortunately there seemed to be very little overlap. The system was organized into user groups, and the hams were normally able to avoid the XXX gang, and vice versa. But whenever I would tell a fellow ham about the system, I'd feel obligated to warn him that on his first visit he might have the feeling that cyberspace's ham radio user group coffee shop had somehow been located in the red light district. "After you download the software, you'll be ready to log on. Now, brace yourself—if you enter the wrong room, you'll find some very weird call signs…"

Internet Phone was very reliable. Using this system, communications was not dependent on the hour of the day or the number of spots on the sun. It ALWAYS worked. There was never any interference or signal fading. It was a useful and interesting bit of technology to have available, but in no way did it pose a threat to short-wave radio. The uncertainties of the ionosphere add important elements of randomness and adventure to communications. For example, years later, in the Azores, one evening I called CQ on what seemed to be a hopelessly dead 15 meter band. A radio amateur in Tennessee responded. We talked for a while, and he told me that he'd once sailed to the Azores. I mentioned that I was the American Consul. He then described the house I was sitting in. He'd been in port on the 4th of July and had been invited to the party that had been held in my house. Think of the odds.

When you are on the short-wave bands, talking to a fellow on the other side of the world, you know that your contact is direct. Unlike the VoIP contacts, the communication is not dependent on any mysterious technological intermediaries. If you are using a radio that you've built yourself, you get even deeper feelings of direct contact, technological independence, and achievement. VoIP is fun and is useful, but it also has no more allure and challenge than a long-distance phone call.

Of course, radio amateurs managed to jazz it up a bit. Soon after Internet Phone became available, articles appeared describing simple systems that would allow radio amateurs to link VoIP to radio systems. In a sense, this allowed us to have the best of both worlds.

In Santo Domingo, I built a system that connected I-Phone to a 145 MHz VHF FM voice transceiver. I set it up so that if anyone called me on I-Phone, the call would be transmitted by the radio. I would be out walking in the park with Elisa. I would have on my belt a walkie-talkie tuned to the frequency of the radio attached to my computer. (Back in the States I'm sure I would have been reluctant to engage in this brazen display of geeky-ness, but when you are in a foreign country, people already think you are a bit weird, so this kind of Public Display of Nerdy-ness is somehow easier.) That little VHF system would normally have a range of only a few miles. It would typically be used for communications with local hams. But now, connected to the internet, it became capable of communications over vast distance. Elisa and I would be walking along, the radio would beep, and I'd take a call from South Africa… or New Zealand… or Japan. I was in geek heaven. This was very cool. Elisa was very patient.

The I-Phone walkie-talkie project had definitely moved me closer to the full-fledged geek zone of propeller beanies, pocket protectors, slide rules, and tape-repaired horned rimmed glasses, but I guess it was the Russian MIR space station that pushed me completely over the edge.

Ham radio on manned space flights started with U.S. astronaut Owen Garriott on Spacelab 3 in 1983, and amateur radio was part of many subsequent Space Shuttle flights. The Russians placed amateur radio equipment on their MIR station in 1990. Antonio and I had from time to time watched MIR fly over head. It was a very big spacecraft, with lots of shiny metal and solar panels, so we could see it quite easily. When, in 1995, U.S. astronaut Norm Thagard became the first American to live onboard MIR, I started getting interested in the possibility of talking to him via ham radio.

Norm Thagard is a very impressive guy. He has a Ph.D. in electrical engineering. He publishes articles on digital and analog electronics circuit design. And, oh yes, he's also a medical doctor, and a decorated Marine Corps jet fighter pilot. With a resume like that, I guess you'd just have to cap it all off by becoming an Astronaut. There'd just be nowhere else to go. But Norm was also one of us.

In his book, *Dragonfly: NASA and the Crisis Aboard MIR*, author Bryan Burrough described Thagard this way:

> "*From his first years in elementary school he was the classic small-town overachiever, fascinated by amateur electronics, fighter planes, and, even before Sputnik, space. He read Asimov and Bradbury and Analog and in seventh grade began writing science fiction stories about space travelers and weird aliens that he sold to other kids for a nickel apiece. In ninth grade he picked up a library book that described how to build a "foxhole radio" and built it himself, earning a ham radio license soon after.*"

A foxhole radio is a kind of crystal radio in which the detector is made out of a razor blade. Norm Thagard's callsign was K4YSY.

The ham radio equipment on MIR was on the two meter band, the same range of frequencies that I'd used for the RS-10 and I-Phone/walkie-talkie adventures. So once again the Yaesu that my dad had bought me went into action.

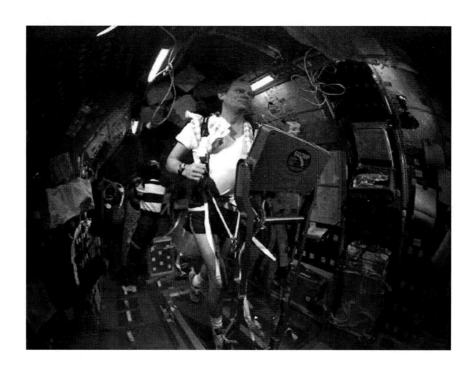

I'd need a special antenna to send my signals to the spacecraft, and *73* magazine had just the thing. *73* had always been the bad boy of ham radio magazines. It was published by Wayne Green, an eccentric maverick who would later have a lot of success with computer magazines. The pages of *73* were filled with odd-ball articles about the electronic projects of radio amateurs around the world. You got the sense that there wasn't a lot of peer review (or any review!) going on in the preparation of the magazine. There was clearly no guarantee that the projects presented would work, or even that they were consistent with the laws of physics. I think that was part of the fun. It was in the pages of *73* that I found the plans for the antenna known as "THE RAY GUN."

This antenna had Buck Rogers written all over it. While my other aerials were harmless wire contraptions with all of the menacing power of unused clothes-lines, this thing looked like a launcher of death rays. Intended to communicate with spacecraft, the Ray Gun looked like it might be able to shoot one down. This thing would shock even those who'd grown accustomed to Pericles' Martian spider roof display. As I made plans to build it, I was thankful that I lived in a country without zoning ordinances, covenants, or neighborhood associations. In suburban America, the Ray Gun would have sent neighbors running for lawyers and court orders, but in the DR, I'd have no problems at all.

The Ray Gun was about six feet long. It was actually two antennas in one. The larger antenna was a five element cubical quad for the 2 meter band. Picture an elongated frame of a box kite. There was a central axis support beam with spreaders that held the five square metallic antenna elements in place. The squares increased in size a bit from the front to the back of the antenna, confirming even to the casual observer that this was a device that would be aimed at something, that this was a device that somehow focused a beam. This, my friends, was indeed some sort of ray gun.

Inside this cubical quad frame was a devilish looking helical antenna for the 70 centimeter band. Picture a bit of DNA wound around the central axis. This antenna was intended for use with sophisticated dual-band, high-orbit satellites. The 2 meter antenna would be for receiving, while the 70 centimeter antenna would be for transmitting. I only needed the 2 meter portion of the device, so I never built the helical portion.

Ideally, satellite antennas should track the spacecraft as it passes over head. To do this you need a complex rotor system, one to turn the antenna on the horizontal axis, and one to raise it above the horizon as necessary. I decided to keep things simple. I would just keep the thing aimed about 45 degrees above the horizon, and then aim it at the appropriate azimuth using a TV antenna rotator controlled from the shack. The "ray" from the Ray Gun wasn't all that narrow, so this arrangement, would, I hoped, usually get enough of my signal up to the MIR.

I visited the lumberyards of Santo Domingo and found some suitable scrap wood that would serve as the central axis and the spreaders. For the actual metallic elements of the antenna, I bought some copper refrigerator tubing. Soon five squares with gradually increasing perimeters were formed, and attached to the spreaders and the central axis. The Ray Gun was ready to be moved in place. As I prepared to put it on prominent display on the roof of my soon-to-be disfigured suburban house, I realized that I'd seen one of these things before. Poindexter, the mad scientist nephew of The Professor in the cartoon series "Felix the Cat", had used something like this. I felt so proud. I knew I was on the right track.

I had a single story house, with a slightly sloping red tile roof. It was actually quite good for my astronomical and electromagnetic avocations. There were a couple of small airshafts that permitted light and air (and occasionally snakes) into interior rooms. For my antenna's support, I simply put a 2X4 down into one of these and secured it with some rope. On top of the 2X4 I placed a small TV antenna rotator. Atop the rotator went the Ray Gun. In the picture you can see the Ray-Gun on the left, the dual-band dipole in the center, and my vertical antenna on the right.

My rooftop antenna farm in Santo Domingo

I don't think I heard any actual gasps from the neighbors, but I kind of felt them. No one actually said anything. Like I said, Santo Domingo is in many ways a very laissez-faire kind of place, where you can do your own thing without worrying about meddling neighbors. And besides, I had diplomatic immunity. Viva El Ray-Gun!

Inside the shack, a very nice little MIR communications station was taking shape. On a small bench I had the computer (to display MIR's position), the Yaesu memorizer two meter transceiver, and the rotator control for the Ray-Gun. Here's what happened:

1 April 1995: I had an almost optimal MIR pass today at 1500 UTC. Approaching from the northwest, MIR rose almost 73 degrees above my horizon. (I'm in Santo Domingo, Dominican Republic.) Best of all, it was 1800 Moscow time and Astronaut Norm Thagard was on the air! I heard Norm work John (W5HUQ) in Jacksonville Florida. Jacksonville is Norm's hometown; I think this gives those of us who are "in the neighborhood" a bit of an advantage! Norm was telling John that he had hoped to take some pictures of Jacksonville, but unfortunately the city was clouded over during this pass. After the MIR was no longer in sight of the continental U.S., I gave Norm a call and got an immediate response. By this time he was almost directly overhead and had a very strong signal. We had a very nice conversation - I told him that I am in the Dominican Republic with the U.S. Embassy; he remarked that while undergoing training for this mission he had been attached to the American Embassy in Moscow. I told him that I'd seen his "shack" the night before on CNN.

So, from my little Santo Domingo radio shack, using junky old equipment and an antenna fashioned from scrap lumber, refrigerator tubing and a TV rotor, I was in voice communication with an American astronaut living on a Russian space station. Poindexter himself would have been impressed.

When something like this happens, you want to tell people about it. But with something like this, you have to be a bit careful. If you work at the American Embassy, it is probably not a good idea to mention over the water cooler that you are staying up late in order to communicate with Russian spaceships that fly over your house at night. You also have to be careful in discussing these operations with potential future in-laws. During this period the local newspaper in Santo Domingo carried a story out of Cuba about an amateur there who had done exactly what I was doing. The article was a reminder of the way in which outsiders would likely react to news of my extraterrestrial contacts. The article appeared under the title "STRANGE OCCURRENCE!" and had a distinctly UFO-ish, X-Files tone to it.

Of course, the amateur radio frequencies provide one very good outlet for this kind of talk, but because of the very specialized nature of this kind of operation even on the geeky ham bands it was hard to find people who could relate to this kind of techie achievement. Internet to the rescue! In an early example of the internet's ability to bring together people around the world with very specialized interests, I started posting reports on my space communication activities on the Compuserve Amateur Satellite forum. Yes, even at this early stage in the development of the internet, a place for "Amateur Satellite" enthusiasts had already been carved out. Soon our network for MIR communicators included Jim, KL7QR, in Alaska; Jan, ZS6BMN, in South Africa; Dave at HZ1AB in Dharhan, Saudi Arabia; Maggie, VK3CFI, and Bill, VK3JT, in Australia. Con, W5BWF, kept up-to-date Keplerian elements (orbital data) available for us and provided reports from Texas. But our most esteemed member was Jay Apt, N5QWL. Jay was in Houston. As a radio amateur, Jay was very much one of the gang. But something set him apart: He was actually going to go to MIR.

Astronaut Jay Apt had already flown on the space shuttle three times. He'd had some real outer-space adventures: During a spacewalk, his glove had been punctured—the hole had sealed up with dried blood. And now he was on Compuserve with us, talking about our contacts with Norm Thagard:

#: 362625 S5/Amateur Satellites

 06-Apr-95 22:42:23

Sb: #362423-#MIR

Fm: Jay Apt 74746,146

To: WILLIAM R. MEARA 74537,1100 (X)

I worked Norm twice 4/6/95 at 13:25 and 15:00 UTC from the Johnson Space Center ARC shack. Also worked him 4/4/95 at 15:15 UTC. I noted that he was not on at the 13:45 UTC pass on 4/4/95. We spoke at length about his activities; he said that he often has holes in his workday schedule, so afternoons and evenings (Moscow / MIR time) are not the only time he is on.

He seemed really happy to be on the air, and above the air! He does request that folks please not interrupt an ongoing QSO (you can hear his side quite well). He'll be up there for a long time, and says that everyone will get a chance. He prefers ragchewing to contest-style "59. Next?" operations. He's working simplex, and really wants to hear the ground side of conversations.

73, Jay N5QWL

07-Apr-95 05:59:07

Fm: WILLIAM R. MEARA 74537,1100

To: Jay Apt 74746,146 (X)

Thanks Jay! I've been able to work Norm twice from Santo Domingo. Quite a thrill! I sent a tape of the QSO to my dad. I'm not normally a mobile operator, but for the last week I've had a two meter rig in the car and I go to work armed with the orbital info for the day. I go out to the parking lot (of the U.S. Embassy) and listen for R0MIR. Send my regards to Norm if you speak to him again.

73 de N2CQR/HI8 Bill

These e-mails from Jay were almost as good as direct contacts with MIR. On April 12 1995, Jay told us that he had again succeeded in contacting the space station, and had taken the opportunity to wish all those on board a "Happy Cosmonautics Day." In his e-mail, he casually mentioned that he had extended these kinds of greetings to MIR once before, but on that occasion he too had been in orbit, in the shuttle Endeavor.

A couple of days later, Jay sent a message that made all of us feel privileged to be in such direct and fraternal contact with him. Like Antonio and I, and many others, Jay would occasionally look up from the radios and watch MIR fly overhead. On April 14, 1995, he sent us this:

14-Apr-95 22:23:46

Sb: MIR view

Fm: Jay Apt 74746,146

To: All

Thanks to Gil Carman's prediction of MIR visible passes, I saw MIR tonight. It was a great feeling for me to be able to watch MIR on the very night I was assigned to a mission to rendezvous with it. It was a lovely sight in the eastern sky, easily visible even with the moon so near full. I stood on my dock and watched it above the water lit up by the moon, double-checking the pass time reading Gil's e-mail by moonlight. Even though it will be a year and a half before I get up there to it, I was moved by the sight.
73, Jay

The Compuserve group was a place to compare notes on contacts, pass around information, brag of achievements, and, occasionally, ask for help. A request for help from South Africa illustrates the global extent and extraterrestrial reach of our cyber-space support group. Jan, ZS6BMN, in South Africa posted a message saying that he had been trying without success to contact MIR. He feared that Norm might not have been aware that there were radio amateurs in that part of the world trying talk to him. Soon after this message, Jay Apt had a contact with Norm. He passed word that the South Africans were listening. Soon there was a triumphant message from Capetown, reporting that Jan had had a very nice contact with MIR. And there was icing on the cake: Thirteen year-old Toby, ZS6BMN, had also managed to talk to Norm.

My own contacts with Norm Thagard were a kind of strange mixture of the routine and the bizarre. On the one hand, they were much like any other contact between radio amateurs. It was friendly banter supported by common interests in technical things, with a dose of fraternity. Elisa would ask me about Norm in exactly the same way that she'd ask about the other fellows who I talked to on the radio. But on the other hand, this guy was in outer space. Sometimes I'd talk to him, and the conversation would be so comfortable and familiar that you'd almost forget this very distinguishing feature. But then something would be said that would make it clear that Norm was in a very different place... In one contact he said he was tired because he'd had a late night. OK, nothing too unusual about that. But then he explained that he'd had to wait up for the rendezvous and docking of a Progress resupply ship. While other guys were waiting up for teenage children, Norm was waiting up for a Progress resupply ship.

My contacts with Norm were really just normal ham radio chit-chat, but out in California, another ham was making a significant contribution to Norm's quality of life. In "Dragonfly," Author Bryan Burrough describes ham radio's contribution of morale support in the early days of Thagard's mission on Mir:

> *For six weeks [Thagard] received no news updates from NASA. Into the breach stepped an American ham radio operator named David Larsen. Larsen, a forty-five year-old psychologist who lived high on a ridge in northern California, had been speaking intermittently with the cosmonauts aboard Mir since the first ham radio was installed aboard the station, in about 1990.... Early in the Mir 18 mission, Thagard seized on Larsen as a reliable channel to slip messages to his wife and others. On Mother's Day, it was Larsen whom Thagard relied upon to send [his wife] flowers. And it was Larsen, not NASA... who on his own sent Thagard regular summaries of world news. 'The news you send, for now, is the only news I get,' Thagard wrote Larsen in an e-mail message.*

My MIR contacts were great fun. One morning I was able to combine visual and radio contact. I told Norm that I'd been able to look up and see MIR flying overhead. He came back with some nice commentary on how the Dominican Republic looked from up above. For a guy who grew up with the space program, and whose interests are astronomy and ham radio, it just doesn't get much better than that.

In July 1995 the Shuttle Atlantis went up and docked with MIR. That was Norm's ride home. After many months in space, he was coming back to earth. In a press conference from Atlantis, he made mention of how ham radio had helped him deal with the psychological challenges of four months in an orbiting craft the size of a Winnebago. That made me feel good—I felt a bit like I'd been part of the support staff.

In Santo Domingo, from the front yard of the house of Elisa's parents, we watched the combined Shuttle-MIR complex go overhead shortly after dusk. (It's amazing what you can see if you know when and where to look.) A few days later I watched TV coverage of the Atlantis touchdown at the Kennedy Space Center. As I watched, I thought of my last radio contact with the MIR. I'd just finished watching the movie version of *The Right Stuff*, Tom Wolfe's wonderful book about project Mercury. On that last radio contact, as MIR slipped close to the horizon and into the static, I couldn't resist using that famous phrase that Mission Control had used to cheer on John Glenn. "Godspeed, Norm Thagard," I transmitted. "Thanks Bill," came the reply from space.

I may have chosen the wrong profession, because I don't really like to travel, and I hate moving. In the months prior to each move, I start getting what I call "my Foreign Service nightmare." In the dream I'm always in the final stages of packing up. Suddenly I notice a door in the house that for some reason I have never bothered to open. When I open it, I discover that behind it there is a beautiful suite of rooms with an amazing ocean view… that I have failed to enjoy during my now-ending stay. No psychologist's couch is needed to figure out this one—it's obviously spurred by the sense of not having seen enough or done enough in the place that I'm now leaving.

My DR versions of this dream started in early 1996; that was the year that I'd be leaving. I had been sent there in 1992 on a two year tour, and I'd extended twice. But now it was time to go.

Dominican post-script:

Jay Apt did fly up to MIR, but he made his trip during the summer I was moving, so I didn't get to talk to him while he was in space.

Pericles passed away a few months after he made his satellite contact. I like to think that our work on the satellites brightened his final days. I have information on him on my web site. His son and his grandson have found their way to the site via Google searches. We had some nice e-mail exchanges.

Gustavo and the Radio Club Dominicano group are still active. I see them when we go back for visits.

Amado the night watchman died a few years after I left. I'm not sure what disease killed him, but I'm fairly sure it was something that wouldn't have been fatal if he'd been living in a developed country. So I guess we can say that he died of third world poverty. His little boy Joel, the kid that used to help him harvest the mangoes and avocados at my house, is now in his late teens and is struggling to get through school. He wants to become a doctor, but third world poverty has a way of crushing ambition. We are still in touch. If there is anyone out there who could help him, please let me know.

I lost contact with Antonio. I hope he is still looking at the stars. Looking back, I regret that I wasn't more generous with him. I should have left that little telescope in the DR; I should have given it to Antonio.

CHAPTER 4
BOATANCHORS IN VIRGINIA
BACK IN THE USA

A young lad of seventeen, known to possess an especially efficient spark, CW, and radiotelephone station, was discovered to be the son of a laboring man in extremely reduced circumstances. The son had attended grammar school until he was able to work, and then he assisted in the support of the family. They were very poor indeed. Yet despite this, the chap had a marvelously complete and efficient station, installed in a miserably small closet in his mother's kitchen. How had he done it? The answer was that he had constructed every last detail of the station himself. Even such complex and intricate structures as headphones and vacuum tubes were homemade! Asked how he had managed to make these products of specialist, he showed the most ingenious construction of headphones from bit of wood and wire. To build the vacuum tubes, he found where a wholesale drug company dumped its broken test tubes, and where the electric light company dumped its burnt out bulbs, and had picked up enough glass to build his own tubes, and enough bits of tungsten wire to make his own filaments. To exhaust the tubes, he made his own mercury vacuum pump from scrap glass. His greatest difficulty was in securing the mercury for his pump. He finally begged enough of this from another amateur. And the tubes were good ones, better than many commercially manufactured and sold. The greatest financial investment that this lad had made was 25 cents for a pair of combination cutting pliers. That is the spirit that has made ham radio.

—From *200 Meters and Down – The Story of Amateur Radio* By Clinton B. DeSoto, published in 1936 by the Amateur Radio Relay League, Inc. West Hartford, Ct. pp 63-64

Moving in the Foreign Service is a bit like getting hit by a tornado. There you are, happily settled in a familiar place. It is a bit weird and foreign, but after four years it definitely has become home. Then the tornado hits. It picks up you, your car, and all of your possessions, and sweeps you out of your house, out of your ham shack. It wipes the antennas off your roof and removes from your house all evidence of your having been there. Then it plops you and your stuff down thousands of miles away.

You might think that going back to your home country would be easy. Wrong. The experts tell us that returning to the USA after living overseas can be even more difficult than an arrival in a foreign country. When you arrive in a place with a name like Tegucigalpa, you brace yourself for differences and you are prepared to go through a period of adjustment. You don't expect to have to do this when going home. But you do. Things have changed while you were away. And you have changed.

I hadn't really been in a supermarket for almost four years. On my first visit to the Falls Church, Virginia "Giant" I noticed that the people behind me in line were giving me unfriendly looks. "Wow, Americans have gotten kind of hostile," I thought. But when I got to the cash register, the girl kind of sighed as she looked at my chock-full shopping cart, and told me that I was in the "Express lane" (ten items or less). Now I understood the hairy eyeballs from my fellow shoppers. In a voice loud enough for other people in the line to hear, I tried to make amends: "I'm sorry, I've been overseas for four years…" I said. "Oh, and I suppose they didn't have supermarkets where you were," replied the gum-chewing cashier. I did myself no good at all by sheepishly replying, "Well, they did, but I had a maid who did the shopping." I got the sense that the people in the line behind me were now looking for trees suitable for an impromptu lynching.

The "Plastic or Paper" question at the check out also caused me confusion. What the heck were they asking me? "Oh, I'll pay with a credit card," I replied. "Plastic or paper?" she asked again. Puzzled, I looked at my credit card and said, "Uh, it's plastic, I think." We went through several rounds of this before I figured out that we were talking about bags, not means of payment. Take dozens of little episodes like this. That's what they call culture shock. Welcome home!

I'd rented a small house in Falls Church, about seven miles West of Washington. My stuff arrived dribs and drabs from the DR, and I launched a half-hearted effort to set up a Virginia version of my radio shack. But it was all very half-hearted, because I was missing (in both senses of the word) the person who had been responsible for my radio re-birth in the Dominican Republic. Nothing was the same without Elisa. She'd been there to cheer me on when I was talking to Norm Thagard. She'd accompanied me on trips to Santo Domingo's "radio row" and to the social occasions (and even an occasional meeting) of the Dominican Radio Club. She'd climbed up on the roof with me to peer through the telescope at whatever celestial marvel that Antonio and I were enthusing about. I missed her terribly. After a couple months of misery I flew down to Santo Domingo and asked her to marry me.

Here's some really good evidence of my wife's selfless tolerance of my geeky tendencies: She agreed that we should include as part of our honeymoon a visit to the Kennedy Space Center in Florida. Looking back on it, I can't believe that I actually came up with that idea. But knowing my wife, I'm not at all surprised that she went along with it.

Once Elisa was there, the Virginia version of the shack started to come together. It was in the basement, but there was a ground level window looking out on our leafy backyard. There were some trees that had excellent dipole support potential.

I quickly found the local radio club. It had a name that harkened back to the earliest days of radio: The Vienna Wireless Society. They met at a community center not far from our house. It was a Northern Virginia version of the Crystal Radio Club and Radio Club Dominicana. Soon I was one of the Vienna Wireless gang. It was fitting that the club had the old-fashioned word "wireless" in its name, because by this point my interest was focused on gear that seemed to have come from the era in which they really did call it "wireless" (and I'm not talking about Wi-Fi).

Back in the Dominican Republic, when I blew the dust off my old Hallicrafters HT-37 and Drake 2-B rig, without realizing it, I was heading into a kind of Rip Van Winkle experience. When I got back on the air after many years away from the hobby, I sort of expected to find other guys on the air using equipment similar to mine. Instead, I found myself surrounded by sophisticated and expensive Japanese-made transceivers. The main manufacturers were Icom, Yaesu, and Kenwood, and on the ham bands, amateurs would talk incessantly about the relative merits of this version or that version of what some guys scornfully called the "rice boxes." I found myself to be a real oddity on the ham bands (and if you are odd there, let's face it guys, you are really odd). I would make contact with someone. After the normal exchange of names, locations, and signal reports, I'd cheerfully say, "Rig here is a Hallicrafters HT-37 with a Drake 2-B."

Many of my contacts didn't seem to know quite how to react to this. Some of them had never heard of the equipment I was using. For them, me saying that I was using a Munchenclobber 654B with a Radonite 6G would have been just as meaningful.

But there was an even more fundamental and revealing misunderstanding going on here. Many of my contacts did not understand why I was telling them about *two* pieces of equipment. They had come into ham radio at a time when transceivers—combination transmitters and receivers—were the norm. My station came from an age when receivers and transmitters were two different devices, usually from two completely different manufacturers. Sometimes I had to explain to the newer hams that such an arrangement was actually possible—no laws of physics were being violated.

Of course, most guys were very gracious about my Rip Van Winkle-ness. They were genuinely interested in my old gear. I think it was refreshing for them to talk to someone whose station was not a carbon copy of all the others they'd talked to recently. Older hams were reminded of their own rigs of yesteryear. Unaware that they were talking to Radio Rip, many of my contacts assumed that I was some sort of nostalgia-driven collector, using old radios for much the same reason that people care for and drive '65 Mustangs. These kinds of comments led me to the Boatanchors sub-culture.

It turns out that my path to old, tube type radios was quite unusual. Most guys who were collecting the older radios and working on them had not, as I had, simply blown the dust off their teenage treasures after years away from the hobby. Most were driven by nostalgia, by fond memories of simpler times and simpler radios. Most had to go out and buy their old gear at auctions or flea markets.

In many ways it is like the attraction of the '65 Mustang. That is a car that the proverbial shade tree mechanic can work on. Like its radio equivalent, it is not stuffed full of complicated microprocessors, and it does not require NASA-like test gear to make it run. The circuitry in the older radios is very simple and easy to understand. These radios were mostly pre-transistor. They were built around the vacuum tube. The Brits call them valves, and that term captures the essence of what they do. They function as valves, with the small signals at the input controlling larger current and voltage variations at the output. They are affectionately called "fire bottles," and—in a reference to the pleasant light that is emitted by their filaments - aficionados proudly proclaim that "REAL RADIOS GLOW IN THE DARK!" Most of these radios were big and heavy, especially in comparison to the lightweight modern gear coming out of Japan. With the aforementioned DX-100 as the archetype, these old tube type radios came to be known as Boatanchors.

The USA is the place to be if you are interested in old radios. The vast majority of the beloved equipment had been churned out by American manufacturers. Hallicrafters, Hammarlund, Heathkit, Collins, and Drake were the main company names. The gear had been built in places like Benton Harbor, Michigan; Mars Hill, North Carolina; and Miamisburg, Ohio. And 1996 was a good year for an anchor-ologist to come home. A lot of guys of my generation were getting hit by the nostalgia wave, and the internet now allowed us to share our interest with other enthusiasts. Boatanchor mailing lists and web sites and USENET groups sprang up. Soon there was an "Old Radio" column in QST, and even a specialist magazine—*Electric Radio*—devoted to older tube-type rigs.

Many of us were now buying equipment that had been beyond our reach as teenagers. In this sense, middle age seemed to have its advantages. eBay soon became the principal market place, but the most fun place to trade gear was the hamfest circuit. For me, this was one of the best parts of my ham radio homecoming.

A hamfest is essentially a flea market for radio fiends. Jean Shepherd wrote that at around age twelve, people come to a crossroads: "one path leads to success, the other to ham radio flea markets." They are always on the weekends, and somehow almost always seem to coincide with inclement weather. They start early. Very early. Oh-dark-thirty early. They are often in community centers or rented halls. You can tell when you are getting close to the site, because the cars around you suddenly have weird antennas and ham radio license plates. On final approach to the parking lot, you'd be guided in by earnest and geeky young men with fluorescent vests and cherished walkie-talkies. All of the people walking in will be carrying empty backpacks or sacks—this is for the soon-to-be acquired junk. Near the entrance, weak coffee and high-cal, pure sugar donuts will be available. It takes energy to get through a hamfest.

Hamfests are almost always divided into two areas: one for the vendors and the other for the flea market/ boot sale. In the vendor area you have to rent a table to hawk your wares. In the boot sale area it's junk from the trunk.

Some of the vendors reminded me of the electronics dealers of Santo Domingo's radio row. Like their Dominican counterparts, they didn't seem to have any real understanding of the stuff they were selling. And they certainly didn't share our rather weird fondness for it. Some of them seemed to have somehow gotten trapped in the hamfest circuit, moving from 'fest to 'fest hawking 6U8A vacuum tubes and other radio-esoterica. I often thought that some of the more eccentric dealers could have found places in Sci-Fi movies of the *Blade Runner* genre (you know, strange people living off the electronic detritus of a technological society).

Hamfests are all about junk. Good junk. Radio junk. Sometimes people will get the idea that the hamfest would be a good place to get people with similar interests together for some socializing. This doesn't really work, because, 1) radio amateurs are not good at socializing (that's why we got into this, remember?), and 2) people go to hamfests hoping to buy some good junk, and if you are standing around socializing, you could be missing out.

Computer gear has been making inroads at the hamfests, but I think most true radio fanatics don't like this. Even though the computer stuff is electronic junk, it is not real radio junk, and it is taking up valuable radio junk space.

The best junk can usually be found in old decaying cardboard boxes underneath the vendors' tables. Boxes of weird variable capacitors. Tubes. Assorted radio knobs. Meters. Boxes of assorted parts that were gathered for projects that apparently never got built. Connectors. Stacks of old radio books and magazines. All these treasures can be found under the tables. Noting that he spends most of his time at hamfests peering into boxes on the floor, one friend of mine suggested that hams put their name tags on the toes of their shoes.

And of course, there were Boatanchors on sale. By the mid-90's the nostalgia wave had greatly increased demand for the older equipment, so for the really choice rigs you had to get to the hamfest even earlier than "normal." That was too early for me, but even at the late-ish hour of 8 am there were usually still some interesting old rigs on sale.

Like most guys, I started buying gear that I'd had as a kid. I picked up a couple of rough looking Heathkit DX-40s—I figured that between them there would be enough good parts for one working transmitter. I did the same thing with a pair of Heathkit DX-60 transmitters. I had my eyes peeled for a Lafayette HA-600A (with jeweled movements!) but did not find one at the hamfests. For this I had to resort to eBay.

Of course, I wanted a Heathkit DX-100, but by the time I got involved these rigs had become highly coveted and hard to find. I had to chuckle when I overheard an on-the-air conversation about DX-100s. It was on the 75 meter band. This band had been the hangout of the young guys that I'd listened to while in High School. Twenty years later many of the same guys were still on the air, on the same band with the same rigs. (You see, ham radio is usually not a passing fad. It is a serious, life-long addiction.) One of the more eccentric regulars was rambling on. He was making what the 75 meter AM crowd called "an old buzzard transmission." This would usually be a monologue of at least 20 minutes or so. It would cover a wide range of very loosely related topics. In the course of one of these transmissions, the amateur noted proudly that he was using a coveted DX-100. He went on to say, "I do have ten of them here… but I don't consider myself a collector."

So I wasn't able to get my hands on an actual DX-100. But I was able to get a small piece of one. Heathkit had taken the variable frequency oscillator (VFO) circuit out of the DX-100, put it in a smaller cabinet and marketed it as their VF-1, an accessory for those transmitters that had been designed for crystal control. The VF-1 had the same eerie green dial as the coveted DX-100. I liked it.

It had the added advantage of not causing hernias. I joked that the VF-1, the DX-40s and the DX-60s combined to become kind of an equivalent of a DX-100. But not really…

My VF-1 was a typical hamfest purchase. When I found it, it looked pretty good on the outside. The price was so low that it wasn't really worth the time or bother to open it up to look at the circuitry. It was cold and raining when I bought it, and Elisa was with me. She is normally very patient, but at this point she was in the early stages of pregnancy, and I felt that I would have been pushing the envelope if I'd insisted on taking the time to open up that VF-1 for a little on-site inspection. So it was only when I got it home that I discovered that this particular Heathkit had been built in a great hurry by a deranged soldering-school drop-out.

That is one of the big variables in the second-hand Heathkit game. The condition of the rig you are buying is determined by the level skill of the guy who'd built it (often on his kitchen table) 25 years earlier. You win some, you lose some. But sometimes even the losers turn out well.

There is a strong element of fraternity in the ranks of ham radio operators, and a strong tradition of helping the other guy out. But at hamfests, this spirit is often overwhelmed (in some people) by the desire to get the highest possible price for their junk. You have to take claims that "It worked fine the last time I plugged it in" with a big grain of salt. But most people are honest, and will let you know if the rig you are eyeing hasn't generated RF since the Great Depression. "Doesn't work you say? Good! I pay extra for that!" was the standard response of one friend. Beyond rewarding the honesty of the seller, he genuinely enjoyed the challenge of getting a seriously defunct old rig un-defuncted.

My VF-1 would have pleased this friend enormously. It was really horrible. It looked like the slipshod assembly had been made worse by periodic bouts of "lets see if this works" circuit modification by someone with little or no knowledge of electronics. It was so bad that I concluded that the only solution was a complete rebuild. And in this project, I came to fully understand the logic behind the "I pay extra for that" position of my hamfest friend.

You can't build Heathkits anymore. The company went out of the ham radio business years ago. Occasionally a complete unbuilt kit from Heath will turn up in someone's basement or attic and will then find its way to eBay, where it will fetch a ridiculously high price. There will usually be a heated debate about whether the new owner should preserve the kit intact ("for historical purposes") or go ahead and have the fun of building it.

That old VF-1 got me as close as I'm probably ever going to get to building a real Heathtkit. I disassembled most of it, and, using a schematic diagram and manual obtained from friends via the internet, began my own second-hand Heathkit building adventure. I got it working, and that green-eyed VF-1 would soon become a part of my evolving boatanchor AM station.

Re-building that old Heathkit was fun, but as I was working on it, something was gnawing at me. I had a score to settle, and this score could only be settled by some genuine, no short-cuts, all-from-scratch homebrewing. I still had to build a receiver. That unfinished teenage project from Congers N.Y. was still bothering me.

As with the homebrew transmitter projects, I didn't rush headlong into the first project I saw. Again, being 39 does (believe it or not) have some advantages over being 15—I'd learned to pick my battles carefully. I took my time in selecting a project. The circuit I eventually chose was a design by Doug DeMaw. Doug was one of the super-stars of ham radio homebrewing. For 18 years he was on the staff of the American Radio Relay League. From there he authored scores of books and articles, all of them aimed at helping amateurs like me build their own radios.

In the June 1982 issue of *QST*, Doug DeMaw had an article entitled, "Build a Barebones CW Superhet." It seemed like it was written just for me. It was perfect. The circuit was very simple, very straightforward. I understood it. I could read the circuit diagram like a map, and understand exactly how the signal made its way from antenna to headphones. As the title indicated, there were no frills, no unnecessary "nice-to-have" circuits that complicated things and increased the possibility of another discouraging receiver failure. I also liked the fact that the receiver was made mostly with discrete components, i.e., with individual transistors, not integrated circuit (IC) chips that contained dozens or hundreds (now millions) of transistors. I was, remember, a radical fundamentalist homebrewer.

UNDERSTANDING: TRANSISTOR AMPLIFIERS

As a kid, I'd tried and failed to understand the inner workings of the transistor. Tubes I could understand, but I found transistors annoyingly mysterious. As fatherhood and concerns about electrocution pushed me into safer low-voltage, solid-state circuitry, I felt obliged to learn how these key devices operated. I felt Shep and Stan and Bollis looking over my shoulder, reminding me that if I didn't understand how the circuitry worked, I'd be just another variety of appliance operator.

Start out by thinking about what we want an amplifier to do. Think about a pipe with a large flow of water through it. Insert a valve. If you set it up right, a small turn of the valve can control the flow of an enormous amount of water. Think of the small turn of the valve handle as the input signal. Think of the large change in the flow of water as the amplified output signal. The flow of water is obviously analogous to the flow of electrons.

Now, replace the water source with a battery. Replace the pipes with wires. And replace the valve with a variable resistor. A small turn on the knob connected to that variable resistor can result in an enormous change in the amount of current flowing through the wires. If you could somehow get a weak signal (say a signal from a distant station coming into the shack from your antenna) to control the value of that variable resistor, you could turn that weak signal into a stronger signal. That's what amplification is all about.

In vacuum tubes, by using a combination of heat and high voltage, a flow of electrons from cathode, through the vacuum, to the anode is set up. Then, a metal grid is placed in the flow. By varying the voltage on that grid, the flow of electrons through the tube can be controlled. The British still, very appropriately, call vacuum tubes "valves."

Transistor amplifiers seek the same kind of control, the same kind of valve action, but they do it in a very different way.

Think of a solid state diode with the battery hooked up so that no current is flowing. This is called reverse bias. The depletion zone is there at the junction, preventing current flow. It is like a layer of insulator separating two conductors. But remember, silicon is a SEMIconductor. Its conductivity can be varied. Shine a laser on that junction, for example, and numerous electrons and holes will be generated. As long as that laser shines on the junction, the zone will no longer be depleted. Current will flow through the diode.

Now, to get a taste of what we are going to try to do with a transistor amplifier, imagine hooking up some circuitry that will allow you to use your voice to vary the strength of the laser beam. You'd be amplitude modulating the beam. Now if you shine the beam on the diode's junction and speak into your microphone, your voice will be—from moment to moment—controlling the number of charge carriers in that junction. Current will flow through the diode, and the current flow will vary with your voice. You'll have a rudimentary amplifier.

This is a key point: Control of the number of charge carriers in a reverse biased diode junction is how most transistor amplifiers work. We don't normally use lasers to exert the control. We use a second diode junction to "emit" charge carriers into the depletion zone, and we let our input signal control the number of charge carriers that are emitted.

Let's start out with a reverse biased N-P junction. There is a big depletion zone. No current is flowing. Now let's add another layer, another bit of N type silicon. We now have an NPN sandwich. The layers are called Collector, Base and Emitter. Remember that N doped silicon is a good conductor. Like copper it has free electrons floating around. The silicon sandwich that we have created is mostly N type material, with a small, thin layer of P in the center. The insertion of this P layer creates two depletion zones that act like insulators. We know the original (Base-Collector) is reverse biased. Let's *forward* bias the new one (Base-Emitter) and see what happens.

We put a negative voltage on the new N layer, the Emitter. As described earlier, the excess electrons from the doping process are pushed into the junction area. The Emitter emits charge carriers, in this case electrons. The middle part of the sandwich is (in this case) a P layer. It is deliberately kept 1) very thin and 2) lightly doped. It is kept thin because we want the two depletion zones to interact. It is kept lightly doped because we don't want the electrons arriving from the new "emitter" layer to simply recombine with holes—we want them to stay as free electrons, thereby increasing the conductivity of what had been the current-blocking depletion layer.

That's where I got stuck when I was trying to understand this as a kid. I was thinking that the only way an electron could be "emitted" from the emitter into the base was by having an electron fall into one of the holes in the base. I couldn't see how this would affect the conductivity of the Base-Collector depletion zone. Indeed, this kind of "recombination" would not improve conductivity. The point I was missing was that most of the electrons that were being sent from the emitter into the base were NOT recombining. They remained free to move into the Base-Collector depletion zone, making that critical area more of a conductor, less of an insulator—more like copper, less like sand.

We eliminate one of the depletion zones by simply forward biasing the Base-Emitter diode. And because the center layer is very thin, the electrons that flow into this bottom depletion zone find themselves attracted by the positive ions that were created in the collector when the electrons from the doping process moved up toward the collector terminal. In effect, they spill over into the top depletion zone. Now the transistor looks like one long piece of N type material. Now our silicon sandwich has become a good conductor, and current can flow from bottom to top.

You can probably see how the arrangement we've just described can be very useful for switching things on and off. Suppose I want to turn a light bulb on using an audio signal (perhaps in a burglar alarm system). I put a battery in series with both a light bulb and the transistor I just described (with connections to the Collector and the Emitter). Perhaps I'll use a diode to convert the audio signal from AC to DC. The DC from the diode is applied to the base. Without any input signal, the depletion zones are like two insulators, preventing any current to flow. But if my audio signal comes in and puts a sufficiently strong DC voltage on the base, the bottom diode in the transistor conducts. Electrons are emitted from the lower N layer (from the emitter), and both depletion zones are transformed from insulators to conductors, current flows from the battery through the light bulb. A small current has controlled a larger current.

I used a circuit very similar to the one just described when I connected my VHF transceiver to the Internet when I was in the Dominican Republic. I needed something to switch the transceiver into transmit mode when the guy at the other end of the internet connection wanted to talk to me. I built a circuit that allowed the audio signal coming out of my sound card to control current flow through a transistor. That current flow in turn controlled a relay that turned the transceiver on and off. This kind of transistorized ON-OFF switch is very important in electronics. A computer is, after all, a vast array of fast switches.

Obviously we are getting close to amplification. Suppose that instead of the ON-OFF devices that we described above, we developed a circuit that would allow an input signal to cause the current in the output circuit to move along with variations in the input signal. Going back to the analogy that we started with, instead of using the transistor as a switch, we'd be using it as a variable resistor. The input signal is usually applied between the base and the emitter. This input signal then controls the number of electrons being emitted into that critically important depletion zone between the base and the collector. As the input signal increases in strength, the number of charge carriers injected into this zone increases, the total resistance through the transistor (from collector to emitter) decreases, and more current flows though the output circuit. If you set things up properly the output signal will be an exact copy of the input signal, only much stronger. It will be an amplifier. Of course, that part about "setting things up properly" involves a lot more understanding—more about this in due course.

If you think you've gained an understanding of how the transistor works, here's a little test. We've been discussing NPN transistors. See if you can sketch out and describe how a PNP transistor functions, describing how the charge carriers inside the device move. Open up that little black box!

Somehow I always found PNP transistors harder to understand than the NPN variety. Maybe it is just because N type material is just more similar to a metallic conductor than is the more exotic, hole-bearing P type stuff. And NPN transistors work by making that troublesome base-collector depletion zone more N-like.

In the PNP transistor, here is what happens: Again, the base-collector junction is reverse biased. When the emitter-base junction is forward biased, electrons in the P type material of the emitter (electrons from the crystal matrix) are pulled away from the junction leaving holes behind and making it appear holes have "flowed" from the terminal toward the junction. At the same time, free electrons in the base are being pushed by the biasing voltage towards the junction. But because the base is very lightly doped, there are not many of these free electrons. Some of these free electrons will jump into the holes of the emitter. This recombination doesn't do us any good. That's why we make the base "lightly doped." We want the holes of the emitter to survive, we don't want them filled by free electrons!

Holes from in the P type material will find themselves close to the ordinary (un-doped) atoms of the base's crystal lattice. Most of those holes near the junction will be positive ions—ordinary lattice atoms short of an electron that went to fill a hole near the positive battery connection. The force of the electric field of the biasing circuit will cause some of the lattice electrons in the N-type base to jump into these needy holes. The key point here is that it is lattice electrons from ordinary silicon atoms (not the free electrons from atoms inserted in the doping process) that are jumping.

When one of these lattice electrons jumps into one of the holes in the emitter, we say that a hole has flowed from the emitter into the base. The emitter has emitted a charge carrier. It has emitted a hole.

The base is very thin. And now you have holes in it. Because of the thinness of the base, these holes sent from the emitter are now close to the reverse biased base-collector junction.

The collector terminal has a negative voltage on it. This negative voltage has pushed electrons down toward the junction. Lattice electrons have been pushed out of their usual places and into the holes closer to the junction. So there are now atoms close to the junction that are negative ions—they have one more electron than they need. And closer to the terminal there are atoms that have become positive ions—they are short an electron. In effect holes have been moved up toward the terminal (but remember, the only things that have really moved are the electrons.)

The electrons that have been pushed down toward the junction are sitting in holes created by the doping process. They are only loosely held in these atoms, because they are not really needed (because of the doping). The atoms that are holding them have become negative ions. These loosely held electrons, already under pressure from the negative voltage of the collector terminal, find themselves very close to newly arrived holes in the base. The electrons in the collector jump into holes in the base, leaving behind holes close to the base-emitter junction.

That base-emitter depletion zone that had been blocking current flow is now gone: You have holes in the collector, holes in the base, and holes in the emitter. The transistor now looks like a piece of conductive P type material. Current flows from the collector through the base into the emitter. And the amount of current flowing from collector to base is controlled by the amount of current flowing from base to emitter. A very small current in the base circuit controls a much larger current in the collector emitter circuit. We have an amplifier.

****

All receivers do essentially the same job: they convert the radio frequency energy that is wiggling the electrons in your antenna to audio frequency energy of sufficient strength to come out of a speaker and wiggle your eardrum.

There are many different ways to do this. A simple crystal radio uses a filtering tuned circuit to select the desired signal. The crystal then converts the radio frequency signal directly into audio frequency and uses the energy of the radio wave itself to drive the earphones. In the oldest versions of this receiver, the crystal is literally a little chunk of crystalline galena (more modern versions use manufactured diodes). You have to poke at this chunk of rock with a phosphor bronze wire (called a "cat's whisker") in an effort to find the "sweet spot." This technique was developed by experiment and trial and error. It was many decades before scientists came to understand what was going on: At the "sweet spot" the galena and the phosphor bronze were forming a PN junction, a crude but effective semiconductor diode. So you see, my discomfort about having to use "store-bought" transistors is not totally unreasonable. The first semiconductors used by radio amateurs were indeed homebrewed. While in Virginia, I bought a chunk of galena and a phosphor bronze cat's whisker and, with headphones on, sat down and poked away at that little rock until I started hearing music in the headphones. This really brings you close to radio's roots. It's amazing that amateurs were making and using their own homebrewed semiconductor diodes long before anyone even knew what a semiconductor was.

Those early, simple crystal radios have in them quite a lot of important electronics-related physics. By taking a look at the crystal radio we can gain insights into the functioning of more complex receiver circuits, and we can begin to see how different electronic components are combined to turn radio signals into sound.

One of the charms of crystal radios is that they do not require a power source: No batteries, no plugs to plug into the wall. They're free! They simply convert the power of the incoming radio wave into sounds in the headphones. But if there is sufficient power in those radio waves for battery-free listening, why not just connect the headphones to the antenna? The strength of the AM signal from the transmitter is varying with the audio from the studio microphones, right? Yes, but we need to do something to that signal to make it audible. We need to "detect" it.

Old books are often very helpful when you are trying to understand radio and electronics. Sometimes the explanations that were given when this subject was new are better and clearer than those offered in more modern texts. One of the oldest radio books on my shelf is entitled *The Principles Underlying Radio Communication*. It is a U.S. Army Signal Corps manual dated May 24, 1921. That means it predates radio broadcasting in the United States. One look at this manual and you can tell that it also predates the kind of mind-numbing routine-ization that would later mark the U.S. Army's approach to radio—for example, opposite the title page there is a very nice and evocative photograph of "Ripples Produced in a Pond by Throwing in a Stone." Obviously the mystery and romance of radio had not yet been squeezed out by the Army bureaucracy.

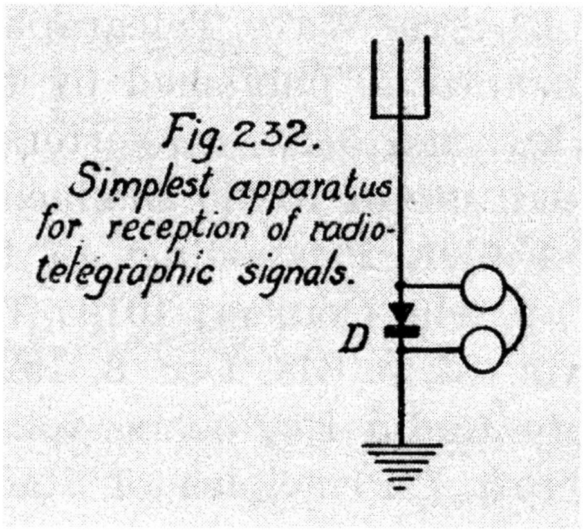

Fig. 232. Simplest apparatus for reception of radio-telegraphic signals.

Page 420 has the diagram for the "Simplest apparatus for reception of radio telegraphic signals." It has just four elements: There is an antenna, a crystal diode, headphones, and a ground connection. Think of the ground as the plate of a very large capacitor. Any voltages induced in the antenna by the incoming radio waves will seek to flow onto this big plate, they will seek to flow to ground.

Page 421 explains why we need the crystal:

"A telephone receiver having magnet windings consisting of a large number of turns of fine wire is a much more sensitive receiving device... The diaphragm can follow the audio frequency variations occurring in ordinary speech, but cannot follow the very rapid radio frequency variations. The effect is as if the diaphragm tried to go both ways at once with the result that no observable motion takes place. For this reason a telephone receiver alone cannot be used to receive radio waves. To remove this difficulty a crystal detector is put in the circuit, which permits current to flow in one direction but not the other; or more exactly, the current in the reverse direction is negligibly small compared with the current in the principle direction."

The crystal detector essentially chops off the bottom half of the signal. Now the sluggish diaphragm in the headphones ("telephone receiver") no longer finds itself being pushed and pulled so quickly that it ends up going nowhere. (Picture a heavyset person being pulled in very rapid succession in opposite directions on each arm—he ends up going nowhere.) But with the bottom half of the signal gone, the diaphragm will be getting tugged in only one direction. Now it will be vibrating, and it will be vibrating at the rate of the audio that the radio wave is carrying.

Imagine a signal coming in from the antenna. When that signal is in the positive portion of the cycle, that crystal diode will appear to be a short circuit to ground, and it is to ground that that part of the signal will go—that part of the signal will not go to the headphones. When the incoming RF signal is in the negative part of the cycle, that crystal diode will look like an open circuit. At that point all of the energy will flow through the headphones. The headphones will be getting hit with a steady stream of pulses of DC—DC pulses that are varying in strength at the rate of the modulating frequency. The inertia of the diaphragm will cause it to not respond to the individual radio frequency pulses, but it will instead move with the audio signal that is superimposed on the RF. And, so, we'll hear voices and music from the headphones.

Today, the crystal radio is explained in a different way: Modern theory teaches that the audio sidebands of an AM signal *mix* with the carrier frequency in the non-linear crystal and produce sum and difference frequencies. The difference frequency is the audio. More on mixers later... This is a good example of how in electronics there is often more than one way to understand something.

More sophisticated receivers make use of a principle of physics and math known as mixing (or multiplicative mixing). For centuries, musicians had known that two different audio tones combine to produce a third tone or "beat frequency." The Italian musician Giuseppe Tartini was the first to describe it, calling it *il Terzo Suono* ("the third sound"). Radio pioneer Reginald Fessenden took this principle and applied it in radio transmitters. Fessenden had coined the word "heterodyne"; it is derived from the Greek heteros (meaning "other"), and dynamis (meaning "force").

The "super-heterodyne" receiver was invented by one of us. Edwin Howard Armstrong grew up in Yonkers, New York, very close to my old stomping grounds.

"From the three windows of his room underneath the cupola on the top floor of his house, Howard had a commanding view of the Hudson and the Palisades; the space and relative distance from the rest of the family offered an excellent place for him to construct radio equipment. By this time the house had electrical light, and Howard suspended two lamps from the ceiling over the worktable so that he might work through the night. His was the highest point in the house the place where he could best send and receive wireless signals. His parents helped him to acquire the necessary wireless paraphernalia—including induction coils, coherers, a telegrapher's key and

earphones, Leyden jars and condensers—and he busied himself constructing and experimenting with electric circuits." (From *"Empire of the Air"* by Tom Lewis.)

Armstrong was a Major in the U.S. Army Signal Corps during World War I. He ran an Army communications lab in Paris. One night while walking home from work, he paused to watch a German bomber attack the city. Thinking of better ways to detect enemy air attacks, Armstrong's thoughts turned to the then difficult problem of how to receive the very high frequency spark signals coming from the electrical system on the plane's engine. The vacuum tubes of the day were unable to work above around 1 MHz, and the sparks from the engines would be at a much higher frequency. Remembering that he had in his laboratory some very stable *low* frequency amplifiers, Armstrong suddenly realized that he could use the heterodyne principal to shift high frequency signals down to the range of his stable low-frequency amplifiers. Thus was born the Superhet.

Modern mixer theory tells us that if you take signals of two different frequencies and pass them through a non-linear device, at the output you will find signals at the sum of the two input frequencies and at the difference of the two frequencies. This is a very useful thing for receiver designers. To make a very simple receiver, you could use a mixer circuit, with energy from the antenna coming into one of the inputs. At the other input, you could put a signal that is almost exactly at the frequency you wish to receive. In this type of receiver, the difference of the two frequencies should be close to the audio frequency signals (the tones of speech or music) that you want to hear. This is called a direct conversion receiver—this is currently a popular receiver circuit for radio amateurs. It was a direct conversion receiver that I tried unsuccessfully to build as a teenager.

Much later, I'd built a simple direct conversion receiver kit from the G-QRP club. It worked as it was supposed to, but it compared very poorly to my beloved Drake 2-B. The Drake was a superheterodyne. That's one of the main reasons I settled on Doug DeMaw's superhet design—I'd been spoiled by the 2-B.

Poor selectivity is the biggest problem of direct conversion receivers—they have difficulty separating the wheat from the chaff. Signals close in frequency to the desired signal will also be audible. Obviously some filtering is needed. It turns out that you can build very good filters if you are designing them for one particular frequency. And it helps if this frequency is low.

So to build a really selective receiver, we start by building a very selective filter for one frequency. Then we use the mixing principle described above to shift the desired incoming signals down to the frequency of our carefully designed filter. Think of the filter as a narrow window. As you tune the receiver, you are actually changing the frequency of the "local oscillator" that feeds the mixer. When the local oscillator is at the frequency that when subtracted from (or added to) the incoming signal produces a mixer product that is at the filter frequency (usually referred to as the "intermediate frequency"), your desired signal goes through the window, and eventually on to your headphones. That is a superheterodyne receiver.

In most superhets, after the filter there is stage called "the product detector." Here the carefully filtered intermediate frequency (IF) signal is mixed with a signal from a "beat frequency oscillator" or BFO. This oscillator is running at, or very close to, the intermediate frequency. In the product detector the BFO signal mixes with the IF signal. The difference or "beat" frequency is at audio frequency levels: either a tone suitable for listening to Morse code, or the varying frequencies of human speech.

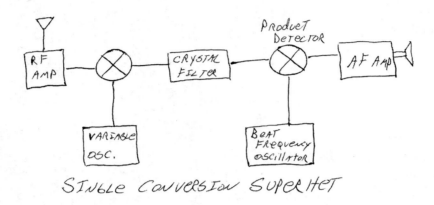

By the way, you might think that the "super" in superhetrodyne is just a bit of advertising promotion, but it is not. Here the super is short for super-sonic (beyond the range of sound). It refers to the fact that the output of that first mixer (the one that feeds the filter) is beyond the range of human hearing. By this logic you could call a Direct Conversion receiver a "Het" but not a "Superhet." But here's another way of looking at it: In the receiver described above, everything to the left of the filter is there to create the SUPERsonic IF. The circuitry to the right of the filter is essentially a direct conversion receiver operating at the IF frequency.

Most of the stages in DeMaw's receiver were built around 40673 Dual Gate MOSFET transistors. MOSFET stands for Metal Oxide Semiconductor Field Effect Transistor. They were obsolete, but I liked them. I found MOSFETs easier to understand than the more common Bipolar Junction Transistors (BJTs) that we have been discussing. A MOSFET basically has a chunk of semiconductor material (N or P) through which a current flows. It enters at the source terminal, and exits at the drain. Along the way it passes a gate made of the opposite kind of semiconductor material. The signal that you need to amplify or process is applied to this gate. Through an effect similar to that of an electric field on a vacuum tube grid, the signal on the gate affects the conductivity between the source and the drain. For me, MOSFETS were reminiscent of vacuum tubes. Tubes I understood, but BJT transistors were still little black boxes for me. And of course, in their simplicity, the MOSFETS were the polar opposite of the very opaque IC black boxes that I was trying to avoid.

The 40673's had, as their name indicated, two gates. This seemed to make them ideally suited for the kind of signal mixing that takes place in a super-heterodyne receiver. I found it relatively easy to visualize two different signals—perhaps one from the antenna and one from the receiver's local oscillator—going into the two gates, each modulating the current flowing from source to drain, and thus, mixing them.

UNDERSTANDING: THE MIXER

That dual gate 40673 was relatively easy to understand, but over the years I frequently became aware of the fact that I didn't really understand what was happening in that mixer circuit. At times I thought that I understood it, but then I'd dig a bit deeper and find that my understanding was incorrect, or at least incomplete. Mixers are absolutely key stages in almost all amateur transmitters and receivers, so I knew that as a radical fundamentalist I'd eventually have to really understand how these circuits work.

There are many paths to confusion in this area. You can be misled by graphical explanations and by "hand waving" verbal descriptions. And I think that purely mathematical explanations fail to provide the kind of intuitive understanding we are looking for. Let me describe some of the pitfalls.

When they get to mixers, some books show three nice graphs of sine waves. They are stacked one over the other. The top two are input signals, each of a different frequency. The third graph is the arithmetic sum of the top two. Moment by moment the signal strengths presented by the top two are added together, and the result is shown on the bottom.

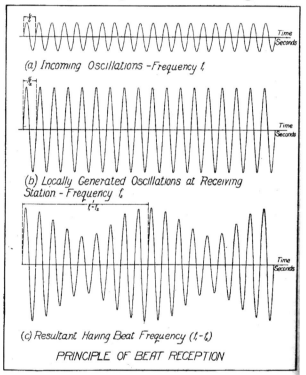

Soon it becomes clear that the shape of the envelope of the resulting graph is varying at a different frequency. A third frequency. What is happening is that because the two input signals are of different frequencies, they are periodically going in and out of phase: one moment both signals are at a positive peak, and they reinforce each other. Later, one is at a positive and the other is at a negative peak; here they cancel each other. It is so simple and easy to see! You can even count the number of cycles that this new signal is going through. And—amazingly—the frequency of this new signal is the arithmetic difference of the first two. Voila! This is *il terzo suono* of Giuseppe Tartini! Suddenly it seems that you understand mixers and superhets. You might think that you now understand how Major Armstrong hoped to convert the electrical engine noise from the German bombers down to a frequency at which he could amplify them.

But you'd be wrong. Sorry about that. It's just not that simple.

It took me a long time to realize that this explanation of mixer action was a kind of children's fable for mixer theory, a misleading fairy tale with sufficient connection to the related field of acoustics to take on an aura of legitimacy.

The first indication that something was amiss came when I looked for the sum frequency. I knew that mixers produce new frequencies at BOTH the difference of the inputs AND at the sum of the inputs. The neat little three-graph presentation seemed to explain the difference frequencies, but what about the sum output? How do we explain that output using these charts? It took me a while to realize that you can't. Because this is not really the explanation of how mixers work.

In the course of writing this book (in 2007), when I got this point I was reminded that my struggle to understand mixers has been a long battle. I was reminded of this when I turned to Google in search for insights. Along with the many learned and highly technical articles that popped up in the search results, I found my own pleas for help going back some ten years. Here is a typical exchange on this subject posted to sci.electonics.basic USENET group in 1997:

> *September 5, 1997 Bill Meara (wme...@erols.com) wrote:*
> *> Here's a nagging little question that has been bothering me for some*
> *> time:*
> *> I have several Physics and Radio books that give very clear*
> *> explanations of how "beat" frequencies are generated in mixer*
> *> circuits. These books have nice little charts showing how the two*
> *> waves combine to produce a third frequency that is the difference*
> *> between the two. Great! Very illuminating.*
> *> But these same books are oddly silent on how the "sum" frequency is*
> *> developed. Can this frequency be explained in a similarly graphic*
> *> manner? Any hints?*
>
> *An excellent question.*
> *It relies on nonlinear circuit elements and high school trigonometry (trig) identities. Ideal mixers have square-law or Vout = Vin^2 characteristics. This means that if you have two signal of different frequencies, vin = s1 + s2 where s1 = cos (2*pi*f1*t) and s2 = cos (2*pi*f2*t), you have vin equal to the sum of two cos, which by trig identity gives vout equal to terms of cos(f1+f1)/2 and cos(f1-f2)/2.*

The response in this exchange is typical of what you get when you ask these kinds of "how do mixers really work" questions. Most experts will immediately come back at you with two things: non-linear elements (like a diode) and trigonometry. The equations seem to be saying that the sum and difference frequencies that we see coming out of mixer circuits are caused by the multiplication of the two input signals.

What? How does that work? At this point many of the books seemed to chicken out on providing non-mathematical explanations. But for me, the math seemed to cry out for some explanation. The equation seemed to be saying that some very simple devices—one diode, for example—are somehow able to take two input signals, multiply them together, and spit out new frequencies that are the arithmetic sum and differences of the two inputs. I found myself thinking, "Diodes are good, but are they *that* good? Who taught them the multiplication tables?" And if we are seeking sum and difference frequencies, why do we eschew addition and subtraction? Why do we use multiplication?

The simple explanation using the three charts and Giuseppe Tartini's *Terzo Suono* explanation kept putting me on the wrong path. I kept coming across examples, mostly from acoustics, that showed two frequencies coming together this way to produce a third frequency. There was, of course, a common experiment in the high school physics lab in which two tuning forks of slightly different frequency are brought together. You can hear the "beat," the difference frequency that results. Where is the "non-linear" element in this case? Displaying what I thought was a somewhat unquestioning acceptance of what they'd learned in engineering school, some folks told me that for this kind of beating to take place there *had to be* a non-linear element. Some suggested that the mixing took place in a non-linear portion of the human ear. Others hinted that the air itself might have non-linear qualities.

This didn't sound right to me. So I built a little circuit that would electrically combine two audio signals. And I would watch the results on my oscilloscope. There'd be no air (or ears) involved. Sure enough, on the scope I could see the beats as the two frequencies came closer together. There were no non-linear diodes doing multiplication. I thought I was getting closer to understanding.

But I wasn't.

The problem with this kind of mixing or combining is that the resulting beat is not "extractable." When I first started seeing that word—extractable—in incoming e-mail messages, I didn't really understand what it meant. Tom Holden, VE3MEO, made it clear: "The beat note that you hear between the two tuning forks is not a new signal—it's just the period between the constructive and destructive interference due to the superposition or addition of the two signals... You can't separate the beat frequency signals from the source signals because subtracting one of the source signals from the waveform leaves you with merely the other. *You can't hear the beat without hearing both forks singing."*

Real mixing is obviously different from this kind of *terzo suono* beating. In a real mixer you want to be able to separate the new frequency from the old ones. You want to be able to extract it so that you can better filter it and amplify it. And you want to leave the input signals behind.

OK, back to the drawing boards.

Back in 1999, I think I kind of came close to a limited understanding of the phenomenon. Here is another USENET exchange:

> *To: ianpur...@integritynet.com.au*
> *Ian: I really like your pages.*
> *I have a question about the theory behind the mixer stage: As was done on your page, explanations of this stage are usually limited to stating that the active device is operated in the non-linear portion of the curve and this results in its operation as a mixer. Given that this is the heart of superheterodyne operation, I've always wished that the explanations would go a bit deeper.*
> *We recently had a very lengthy and interesting discussion on this in the sci.electronics.basics newsgroup. I came to some conclusions about mixer operation that (I hope) may provide the kind of explanation that I think is needed for beginners and non-engineers to understand mixers:*
> *--When we say that the active device is operated in the non-linear portion of its operating curve, we are really saying that we are biasing it so that each of the two input signals will—in effect—vary the amount of amplification that the other receives from the device.*
> *-- When this happens, the output of the device is a waveform that contains sum and difference frequencies. If we ask WHY this happens, we have to be satisfied with an answer that points to mathematics: If you combine two signals in the manner described above, mathematical principles dictate that the resulting waveform contains sum and difference frequencies.*
> *Please let me know what you think of this explanation—does it make sense, is it consistent with accepted theory?*
> *Again, thanks for the great web site. I will be visiting often.*
> *73 Bill N2CQR*
>
> *Bill Meara <wme...@erols.com> wrote: receives from the device.*
> *Excellent. This is a very good definition of non-linearity. Sometimes "amplification" isn't involved, as when we use a diode mixer, but in that*

case each signal varies the amount of _attenuation_ that the other receives from the device. I like it, and I think that the explanation is about as useful as any.
For a mathematical analysis, you might want to consider that a non-linear mixer actually _multiplies_ the two signals, rather than adding them. I think I've got this right, anyway...
M Kinsler

So, back in 1999 I seem to have sort of accepted that if you take two signals of different frequency and feed them into a non-linear device, the math tells us that in the output you will get sum and difference frequencies. I also seem to have been coming close to understanding the need for non-linearity: in order for the signals to really "mix" one signal has to affect how the other signal passes through the device. My thinking was that if you have one signal in effect varying the bias on a transistor as that signal goes through its cycle, another signal going through that device will see the device as being extremely non-linear. It will get mixed with the first signal. (In retrospect, my understanding of the role of non-linearity was still quite flaky.)

Still, I was not satisfied with my understanding of the mixers. I thought it was a bit of a cop-out to just say, "Well, the math tells us that if you multiply two sine waves, the output will contain sum and difference products." Math-oriented scientists and engineers often pour scorn on what they call "arm waving" non-mathematical descriptions. But I think there is some room for scorn in the opposite direction: I don't think that memorizing a trig formula means that you really understand how a mixer works. In "Empire of the Air," Tom Lewis writes: "At Columbia, [Edwin Howard] Armstrong developed another trait that displeased some of the staff and would annoy others later in life: his distrust of mathematical explanations to account for phenomena of the physical world. All too often he found his professors taking refuge in such abstractions when faced with a difficult and seemingly intractable conundrum... Time and again as an undergraduate at Columbia, Armstrong had refused to seek in mathematics a refuge from physical realities."

I guess I still yearned for the clarity and intuitive understanding that had been (falsely) promised by those three nice beat frequency charts. Time and time again, as I dug into old textbooks and ARRL Handbooks and promising web sites served up by Google, I was disappointed.

Then I found it.

It was in the Summer 1999 issue of SPRAT, the quarterly journal of the G-QRP Club. Leon Williams, VK2DOB, of Australia had written an article entitled "CMOS Mixer Experiments." In it he wrote, "Generally, mixer theory is explained with the use of complicated maths, but with switching type mixers it can be very intuitive to study them with simple waveform diagrams."

Eureka! Finally I had found someone else who was dissatisfied with trigonometry, someone else who yearned for the clarity of diagrams. Leon's article had waveform diagrams that showed, clearly, BOTH sum and difference output frequencies.

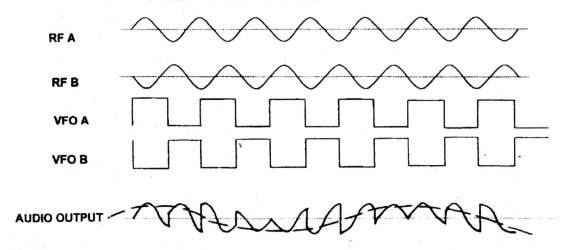

Switching mixers apply the same principles used in other kinds of mixers. As the name implies, they switch the mixing device on and off. This is non-linearity in the extreme.

Not all mixers operate this way. In non-switching mixers the device is not switched on and off, instead one of the signals varies the amount of gain or attenuation that the other signal will face. And (as we will see) it does this in a non-linear way. But the basic principles are the same in both switching and non-switching mixers, and as Leon points out, the switching circuits provide an opportunity for an intuitive understanding of how mixers work.

Let's take a look at Leon's circuit. On the left we have a signal coming in from the antenna. It goes through a transformer and is then applied to two gate devices. Pins 5 and 13 of these gates determine whether the signals at pins 4 and 1 will be passed on to pins 3 and 2 respectively. Whenever there is a positive signal on gate 5 or on gate 13, signals on those gaps can pass through the device. If there is no positive signal on these gates, no signals pass. Don't worry about pins 6-12.

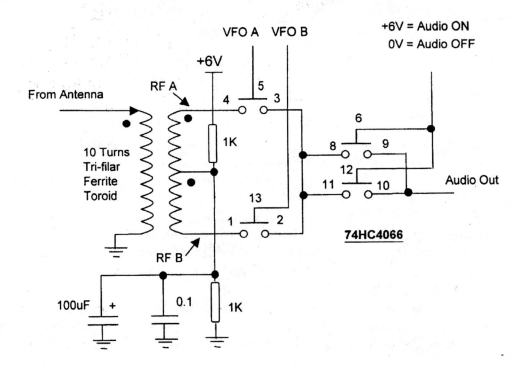

RF A is the signal going to pin 4, RF B is the "flip side" of the same signal going to pin 1. VFO A is a square wave Variable Frequency Oscillator signal at Pin 5. It is going from zero to some positive voltage. VFO B is the flip side. It too goes from zero to some positive voltage.

Look at the schematic. Imagine pins 5 and 13 descending to bridge the gaps whenever they are given a positive voltage. That square wave signal from the VFO is going to chop up that signal coming in from the antenna. It is the result of this chopping that gives us the sum and difference frequencies. Take a ruler, place it vertically across the waveforms, and follow the progress of the VFO and RF signals as they mix in the gates. You will see that whenever pin 5 is positive, the RF signal that is on pin 4 at that moment will be passed to the output. The same process takes place on the lower gate. The results show up on the bottom "AUDIO OUTPUT" curve.

Now, count up the number of cycles in the RF, and the number of cycles in the VFO. Take a look at the output. You will find that that long lazy curve traces the overall rise and fall of the output signal. You will notice that its frequency equals RF frequency minus VFO frequency. Count up the number of peaks in the choppy wave form contained within that lazy curve. You will find that that equals RF frequency plus VFO frequency.

Thanks Leon!

Back to the math for a second. Why do they say that those diodes multiply? And what do trigonometric sines have to do with all this?

First the sines. Most of the signals we are dealing with are the result of some sort of circular or oscillating motion—coils that are being spun around magnets, resonant circuits that behave like a playground swing. For this reason, the trigonometry of circles can be used to determine the amplitude of a signal at any given instant. Take the peak value of a sine wave signal, and multiply it by sin[2π(freq)(time)] and you will get the instantaneous value of that signal.

When we say that mixers multiply, it is important to realize that we are NOT saying they multiply frequencies. We are saying they multiply the instantaneous amplitudes of the input signals. And it is that multiplication that results in the generation of the sum and difference frequencies.

Why multiplication? Again, by looking at switching mixers, it is easier to understand. Consider one of the gates in Leon's mixer. The RF input is a sine wave. Its instantaneous value varies according to Peak*sin(2πft). The VFO signal is either positive or 0. If it is positive, the RF signal passes through the gate. We can say that if it is on, it will have a value of 1. If it is 0, no signal passes through the gate. Mathematically we can say that the output is a multiplication product of the two inputs. If RF is at 1.2, and VFO is positive (1), the output from that gate will be 1.2x1=1.2. If RF is at 1.2 and VFO is at 0, the output will be 1.2x0=0

Note that this is very different from the simple summation in the "children's fable" presented at the beginning. If addition were at work here, we'd expect the outputs to look like 1.2+1=2.2 or 1.2+0=1.2 But that is clearly NOT what we'd get with a switching mixer. Clearly multiplication is the operation that best models this circuit.

Now this doesn't mean that in every mixer circuit one input with an instantaneous input of 2 volts and another with an instantaneous input of 3 volts will result in an instantaneous output of 6 volts. After all, some mixers are made up of transistors that are capable of amplification, but others use simple diodes, and these diodes can't amplify. Different mixing circuits use different kinds of devices, different input levels, and different biasing voltages. So the outputs will vary quite a bit. But the shape of the output waveform will resemble the waveform that results when you multiply the instantaneous values of two input waveforms. We can say that in addition to the multiplication that is the heart of the process, there are also mathematical constants and offsets that result from the particular characteristics of individual circuits.

You can use a simple spreadsheet program to get a feel for this. Set up two columns each with the formula Peak*sin(2πft). Assign different values of frequency to each column. Set up another column for time—make it 1-100 and think of each division as a block of time. Graph the results. Then run a third column that multiplies the first two. And put this third column on the same graph. You'll see the mixer action.

Leon's switching mixer circuit helped me get a bit more of the kind of intuitive understanding that I'm always looking for. Later on, through a more careful reading of *Experimental Methods in RF Design*'s mixer chapter, I think I started to understand how non-switching mixers work, and why non-linearity is an essential element of a mixer circuit.

Jean Baptiste Joseph Fourier (1768-1830) discovered that any complex periodic waveform can be shown to be the result of the combination of a set of sine waves of different frequencies. Here's a great illustration of this principle. It is from ON7YD's web site. The darkest line is the complex signal that results from the sine waves that are shown around it. A picture is worth a thousand words.

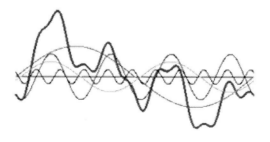

The key idea here is that if you see a complex periodic (repeating) waveform, you should realize that "beneath" that waveform, there are a number of nice clean sine waves. And here is where non-linearity as an essential element in mixing comes in.

Let's consider two devices, both with dual inputs. One is set up to be very linear. The other is set up to be non-linear. Let's put two signals of different frequencies into each input. The first input is 1 volt peak at 1 MHz, the second input is .1 volt peak at 10 MHz.

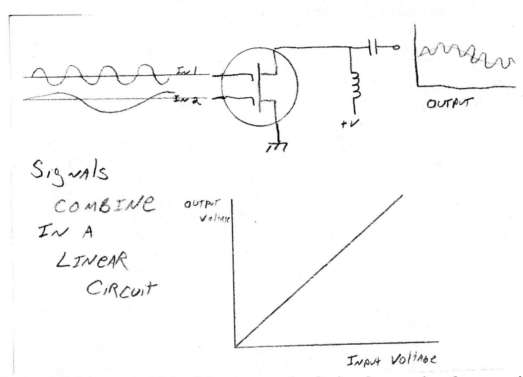

Signals combine in a linear circuit

In the linear circuit, we can think of the stronger 1 volt signal as moving the operating point of the device up and down, up and down along the very straight line that describes the relationship between input and output in this circuit. As it does so, the weaker 10 MHz signal just sort of rides along.

If we look at the output we can clearly see the two signals, one riding along with the other. The output waveform is not complicated, and it seems clear that there are only two signals that you could get out of that via filtering: the two input signals. This is just like the acoustic situation that caused me so much confusion. The key thing to remember here is that the two signals are not really mixing.

Now let's look at the non-linear circuit. Now the operating curve really is curved. The weaker 10 MHz signal will once again, in a sense, be riding along on the stronger 1 MHz signal, but that 1 MHz signal is no longer moving up and down on that nice straight line. Now it is on that curve. Now the two signals really "mix", mixing almost to the same extent that liquids of two different colors mix in a blender. You can see how the curved operating characteristic—the non-linearity— causes the two signals to mix.

Out of the non-linear circuit a very complex periodic waveform emerges. It is a complicated mess, but Fourier tells us that any complex periodic waveform can be seen as being composed of sine waves of many different frequencies. If we were to dissect this output waveform of this device, we'd find the two original signals, harmonics of these signals, and, most importantly, new signals at the sum and difference frequencies of the input frequencies. And this complex signal CAN be dissected. To do this, we make use of "balanced" devices to cancel out the input signals, and filters to shave away the harmonics and perhaps either the sum or the difference output. We can set things up so that only one frequency emerges from the mix. That is extremely useful.

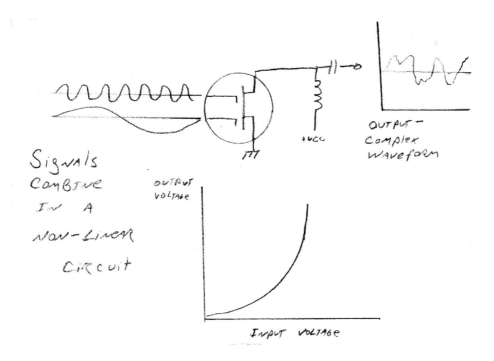

I think (hope!) I've made progress in my effort to understand mixing; I think I've moved far beyond both hand waving acoustics-based fairy tales, and the almost equally unsatisfactory approach that equates understanding with the ability to regurgitate trig formulas. I now understand the difference between mixing and adding. I know why multiplication (and not addition) is the math operation that describes what happens in a mixer. Most importantly I think, I now know *why* you need a non-linear device to have true mixing. Fourier provides the answer: That bend in the operating curve of a non-linear device causes the output to be the kind of complex periodic waveform that contains many different sine waves. And among those waves are sum and difference frequencies.

Now I must admit that *how* it is that among those sine waves there are the exact sum and difference frequencies of the inputs, well, for me that remains a bit of a mystery. But it kind of makes sense…

****

By the time I found it, DeMaw's "Barebones Superhet" design was already fifteen years-old, and 40673's were getting hard-to-find. Here the fraternity of solder melters came to my rescue. I mentioned my parts problems on the rec.radio.amateur.homebrew USENET group, and a fellow ham came to my rescue. Soon a small package arrived (free of charge) with a batch of the needed parts.

I decided to be very careful and methodical in building this receiver. I would break the circuit up into individual stages, then build and test each stage separately. Only when I knew that all of them were working well would I put them all together.

Our lives were getting pretty busy at this point. I was working at State Department headquarters and Elisa was coping not only with pregnancy, but also with the challenges of life in a new country and language. So time management became important; I used the early morning hours and my lunch breaks to work on this project.

In his receiver, Doug DeMaw had used etched printed circuit (PC) boards. Wanting to leave nothing to chance, I decided to follow his lead. A special, factory made board was available for this receiver, but as a radical fundamentalist homebrewer, I of course rejected this option. I joked to myself that I wanted to build a receiver with Politically Correct PC boards—PC PC boards. I would have to design and build my own. The design of the boards became one of my lunch hour tasks.

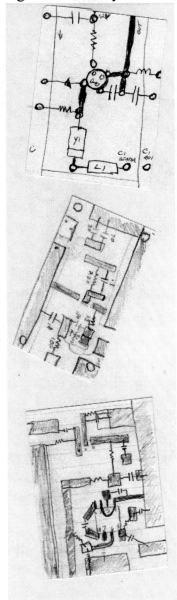

Each day I would head to work with some index cards, a pencil, and the schematic diagram for one stage of the receiver in my shirt pocket. After a morning of dealing with U.S.—Honduran relations, I'd eat my sandwich and head for the Department of State's Thomas Jefferson library. This is one of the nation's oldest libraries—Jefferson had served as one of our first diplomats, and later as Secretary of State, and his collection of books had given the library its start. I told no one at work what I was doing—revealing that you are spending your lunch hour in a corner of the library designing the circuit board layout for a variable crystal oscillator would not be a career-enhancing move for a diplomat. But sometimes I'd look up at the library's statue of Thomas Jefferson. Knowing of his deep interest in science and technology, I thought that he would approve of my semi-clandestine mid-day design work.

I had to convert the electrical design on the schematic to a physical layout on the board. I had to consider the approximate size of each component, and figure out where each part would be placed. I had to keep some principals of physics in mind when doing this—I had to make sure that the output circuits of amplifier stages were kept as far as possible from the inputs—there is an old adage in electronic homebrewing: When you start building an amplifier, you often end up with an oscillator. Putting inputs close to outputs is a good way to get this unhappy result. Once the layout was done, I'd draw a pattern for the etching of the board—which portions of the copper would be etched away, which portions would remain to serve as conductors.

When you tell other radio amateurs that you are homebrewing equipment, a very common response is, "Wow, I'd do that too, if I only had the time!" I don't like that response. Sometimes it is sincere, but often I think it is the defensive response of someone not comfortable with his appliance operator status. This kind of comment often seems to give insufficient recognition to the skill involved in getting a homebrew rig working—it seems to imply that time is the only necessary ingredient. There is also an implication that the homebrewer must have an unimportant job that allows him time for silly tinkering. Or that perhaps he was neglecting his family. The fellow who makes the "I don't have time" remark will then sometimes go on to tell you that his real interest in ham radio is contesting (an activity that typically requires participants to engage in 48 hour, all-weekend-long marathons of on-the-air contact making). Or he'll make these comments while loading his golf clubs in the trunk of his car. Modern life puts time constraints on all of us, but I think my lunchtime design sessions show that when there's a will, there's a way. And believe me, those circuit boards provided good recreation—in the full sense of the word. After lunch I'd return to the problems of Honduras refreshed and better able to do my job.

Most of the design and planning was done during the lunch hours in the library; actual construction would take place at home on Saturday and Sunday mornings. There are many ways to etch a circuit board. I was at the simple and unsophisticated end of the technique spectrum. To cover the areas where I wanted the copper to remain, I used small pieces of ordinary electrician's tape. For the etchant, I used a RadioShack product.

I was doing about one board per week. Seven boards would hold all of the circuitry. I took my time. Each circuit board would be etched and then the needed holes drilled. Parts would be carefully soldered in place. My rudimentary test gear allowed me to see if the oscillators were oscillating and if the amplifiers were amplifying.

After about six weeks of this, most of the boards were fully populated with parts. I knew that each one of them worked individually, but I hadn't yet tried putting them all together. One Saturday morning, using test leads, I temporarily connected inputs to outputs and applied power. I sniffed around for the smell of burning electronics—there was none; I'd passed the smoke test. I was still in the process of checking my improvised wiring when I heard a very faint Morse code signal. At first I thought that it was coming from the Drake 2-B across the room, but the Drake was not even on. The only powered-up receiver in that room was the haywire creation on my work bench. Suddenly, that collection of parts and PC boards had become a receiver.

Doug DeMaw's receiver was designed for CW—for Morse code. CW signals are very narrow. They use very little spectrum. This is the result of a property of physics described in Claude Shannon's Information Theory: As you increase the rate at which information is transmitted, the width of the channel you are using needs to increase to ensure error free transmission. The information transmission rate is relatively low for Morse Code signals—you are sending text at around 20 words per minute (if you are good!). So you can use narrow filters in a CW receiver. This is advantageous because with the narrow filters you will not only be able to block out potentially interfering signals that are close in frequency to yours, but you will also be able to reduce the amount of noise (static) going through the receiver.

On the morning of that first test, I had not yet completed the filter. Doug's design had called for the use of three "color burst" crystals. These are devices mass-produced for color TV sets. Their frequency is 3.579 MHz. This was the intermediate frequency (IF) of my receiver—the narrow window that would give it selectivity. But with only one crystal in the temporary filter circuit, the window was almost wide open. The receiver was valiantly picking up signals, but it was picking up many of them, and it was unable to separate the wheat from the chaff. But on that first day of testing, this lack of selectivity turned out to be an advantage. As late afternoon rolled around, I decided to tune around a bit with my new creation. All day I'd had it tuned to the lower, CW portions of the 20 meter band. But now I was tuning up higher, to the portion of the band where telegraph keys were replaced by microphones.

Around 14.200 MHz, a strong signal carrying a melodious voice appeared. It was EA3OT—"Echo Alpha Three OOOOOLD Timer." Phone signals can sometimes sound a bit muffled and constrained by the filters of commercial receivers, but my little creation was at this point as wide as a barn door, and it was letting Miguel's signal come through in all its glory. It sounded like he was sitting in the room with me. I sat there for a moment, marveling at the miracle of radio as I listened to that signal from far-off Barcelona. I recalled a phrase used by one of the guys on one of the internet mailing lists: I was listening to the magic that you can only hear with a receiver that you have built yourself.

Of course, there were some problems. This is not "plug and play" radio. There was some debugging to do.

The audio amplifier that drove the speaker would (following the old maxim) convert itself into a screaming audio oscillator if I turned up the volume a bit too high. So I had to tame the beast.

The "local" oscillator that provided the RF energy to convert the incoming 14 MHz signals down to 3.579 MHz sometimes didn't start quickly. I had to change the values of some of the feedback capacitors to get it to start properly.

There was a very strong religious short wave broadcast station not far in frequency from the tuning range of my receiver. The simple filters at the "front end" were no match for this electromagnetic giant, and as I tried to listen to Morse code conversations, in the background I could hear fire and brimstone.

The 3.579 MHz intermediate frequency of this receiver was smack in the middle of the 80 meter CW band. So in addition to the desired 20 meter signals, strong signals close to the intermediate frequency were making it through. It just so happens that the venerable ARRL Headquarters station W1AW (that I'd listened to and visited as a kid), had its daily Morse code practice sessions very close to 3.579 MHz. So when the preacher wasn't yelling at me, W1AW seemed to be exhorting me to improve my code speed. All this required me to beef up the front end filtering to keep the undesired signals out. I got rid of both W1AW and the religious broadcasters; my code speed may have suffered, but I don't think I've become any more sinful.

There were mechanical problems too. Like most superhets, this receiver was tuned by changing the frequency of a Variable Frequency Oscillator. In my case it was actually a Variable Crystal Oscillator—a kind of hybrid circuit in which a very stable crystal determines the frequency of oscillation, but with some additional components that allowed the frequency to be moved around a bit. A front panel knob was connected to the variable capacitor that did the actual shifting. A variable capacitor is a collection of small metal plates. You vary the capacitance by meshing and un-meshing rotor plates with stator plates.

The variable capacitor that I was using had come out of an old Swan 240 transceiver that Pericles had given to me. I liked including this part in the receiver—not only was it just what I needed from the technical standpoint, but putting it to use made me feel like I was including my old friend in this new project. Such are the satisfactions that can come from scratch-building with junk box parts.

The Swan capacitor had a "reduction drive" that slowed and smoothed the tuning. This was nice, but it made it impossible for me to tell from looking at the front panel what frequency I was on. In the early days I'd have to sit up and peer over the front panel to see the configuration of rotor and stator blades on the tuning capacitor. This was clearly not the most comfortable way to read the frequency. So I started looking around for something that I could attach to the part of the shaft mechanism that actually moved with the rotors. In these situations, ham radio operators often turn to the kitchen for materials. We still use the term "bread-boarding" for the construction of a temporary prototype circuit. This comes from the early practice of using an actual breadboard (often purloined from mom's kitchen) as a base on which to build a radio. Numerous clothes pins, clothes lines, Tupperware jars, reams of aluminum foil, and yes, a few breadboards, have been appropriated for my own radio experimentation. In this case, a coffee can top proved to be just what was needed. That metal disc that you throw away after opening a can of Maxwell House (appropriate brand, don't you think?) fit perfectly. I drilled out a center hole for the control knob, and four smaller holes for the screws that would connect the disk to a small skirt on the rotor blade shaft. A small triangular bit of electrical tape was placed just above the disk on the front panel, and I marked the coffee can top with the tuning frequencies. Done. By modern standards, this was a very unsophisticated frequency readout mechanism. There were no glowing numerals out to multiple decimal places. But I love that simple little readout. It is *my* innovation, my own solution to a technical problem. I wouldn't trade it for all the glowing numerals in the world.

De-bugging complete, I paired up my new Superhet with the transmitter that I'd built in the Dominican Republic, and went on the air and made my first completely homebrew QSO. Shep, Stan, Bollis, Vlad, and Hilmar would have been proud.

One of the nicest things about this project was that the designer himself found out that someone was making use of his old design. My frequent pleas for help on the homebrewer USENET group came to the attention of Doug DeMaw's son. He passed my messages on to his dad. Doug sent me some nice messages, letting me know that he was happy to learn that after so many years, someone was still building his little Barbados Barebone Superhet. I didn't realize that Doug was very ill when he wrote to me. He passed away not long after our e-mail exchange.

I had a couple of other encounters with Barebones receivers. In response to my e-mailed pleas for help, Michael Hopkins, AB5L, sent me a partially completed kit for this project. Michael was the author of a very entertaining series of stories about the re-incarnation of amateur radio homebrew legend Frank Jones; in the stories, Frank comes back to life and forms a militia-like group dedicated to winning back for radio amateurs the old 5 meter band (currently used by baby monitors). Michael's kit provided many good parts for other projects, and one of these days I may stuff the PC board with parts in order to have yet another Barebones Superhet.

Then another one came my way. Someone on the internet was selling a completed kit. I bought it, and put it away for another day. Years later, I started working on it. The design had been modified by the original builder: The IF frequency had been shifted a bit (to take advantage of available crystals) and the variable crystal oscillator had been turned into a varactor-diode tuned Variable Frequency Oscillator. As I tried to get the receiver going, I started asking questions on the QRP-L mailing list. Dale Parfitt, W4OP, came to my rescue and we started discussing the technical problems. After a few exchanges, Dale said the receiver sounded familiar. It was only then that we realized that the receiver I was working on had been built... by Dale Parfitt!

Obviously I had a lot of fun with those Barebones receivers.

Another good thing about coming back to the States was that I could now get involved in AM radio, in the kind of 75 meter contacts that I had listened to as a kid. Northern Virginia was close to the New England epicenter of this activity, and AM radio was a natural complement to my interest in Boatanchors.

At first I pressed my trusty HT-37 into service and started to talk to AM operators on the 75 meter and 40 meter bands. It was great fun. There really is something special about AM contacts, there is a warmth to them. I think this comes from the sense of presence you get from the carrier. When you are using single sideband, if the operator stops talking, his transmitter is not emitting a signal—during the natural pauses in conversation you can hear background static—it's as if he's disappeared. With AM, the carrier is always there. It seems to block out the static and noise, and it creates this feeling of presence in the contact. The other guy throws the switch to transmit. You hear a nice Ker-chunk from the many large relays in his transmitter, and the static disappears. Perhaps you'll hear the rustle of papers being moved around on his desk, or the sound of him adjusting his chair, or putting down his coffee cup. Then he speaks. The carrier creates an entirely different feeling for AM QSOs—it feels like you are sitting around a campfire, or perhaps around the kitchen table in a friend's house.

UNDERSTANDING: MODULATION, AM, DSB, SSB

Everyone knows what AM is, right? Just about every broadcast band radio in the world has that little AM-FM switch on it, and it is widely known that in AM we vary the AMPLITUDE of the signal while in FM we vary the FREQUENCY. Simple enough, right? Well, as with most things in electronics, here things are not as simple as they might seem.

As with the descriptions of mixers, the presentation of AM modulation in the electronics books (especially those for beginners) is often a simplified version of reality, a kind of radio-theory children's story, sort of an electromagnetic fable. They usually included a diagram showing the steady carrier, with no modulation. It would be at radio frequency, let's say 3885 kilohertz. Then, an audio signal would be introduced via a circuit called a modulator. The strength of the RF signal would then be shown to be varying along with the audio signal.

At a certain level, that is what happens, but in amateur radio we quickly move beyond the point where this kind of simple presentation is useful. In electronics, there are often many layers of complexity lying underneath the simple explanations presented in the handbooks. It's like the picture on a TV screen: First you just see the picture, but look closer and you see the pixels. Look closer still and there are electrons hitting the screen...

To get to a deeper understanding of modulation, start out by understanding that modulation is just another version of mixing. To create our AM signal we mix the carrier (RF) signal with audio from our microphone and AF amplifiers. At the output we have the carrier, the audio, and

the sum and difference frequencies. Usually in an AM transmitter, this mixing takes place in the final amplifier stage. The audio signal is easily filtered out, and what remains are the carrier signal, and the sum and difference frequencies. We call the sum frequencies the upper sideband and the difference frequencies the lower sideband.

Suppose you have our carrier at 3885 kilocycles. You whistle into the mic (as you do—admit it!) and produce a nice 1 kilocycle audio tone. At this point your transmitter will have a strong carrier at 3885 kilocycles, and two weaker signals, one at 3884 kilocycles, and one at 3886 kilocycles. If you stop whistling (please do!) and start talking, your voice will be producing a range of frequencies. The audio amplifiers can be set up to respond to signals from around 300 to around 3000 kilocycles, so as you speak you will be producing upper and lower sidebands stretching above and below the carrier frequency.

The carrier stays steady, but the energy in these sidebands rises and falls with your voice. If you consider the total energy in the entire range of frequencies from 3882 to 3888 kilocycles, the old, simplistic description of AM signals varying in amplitude will be quite true. But if you take a closer look at what is happening within that frequency range (you can do this with a spectrum analyzer) you will see that the carrier really remains steady as the sidebands vary in strength and frequency with the operator's voice.

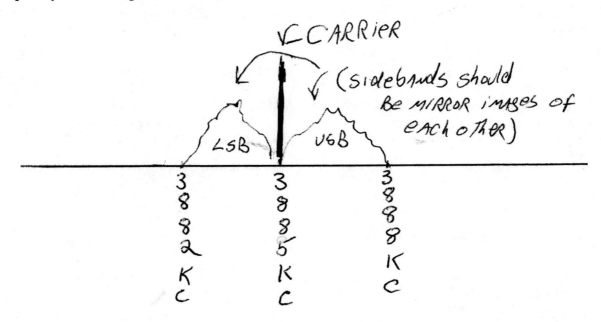

What about all those diagrams showing the "modulation envelope" with the strength of the signal varying with the strength of the audio input? Was that just another electromagnetic fairy tale? No. There is truth in those diagrams, it's just a bit more complicated than it might seem. I had to dig into this theory after I realized that I didn't understand how my good-old Hallicrafters HT-37 transmitter produced single-sideband signals.

Let's stick with ordinary AM for a minute. At the output of the transmitter you have three different signals of three slightly different frequencies. There is the carrier, lets say it is at 3.8 MHz. Assume a 1 kHz tone is being fed into the microphone (you are whistling again!). So now there is also an upper sideband signal at 3.801 MHz and a lower sideband signal at 3.799 MHz.

Think back to the heterodyning of Terzi's *Terzo Suono*. Think about how the different frequencies came in and out of phase. Something similar happens here.

Let's pick a reference moment. At some point, each of the three signals will be more or less in phase. They will be hitting positive and negative peaks in their cycles more or less at the same time. This will correspond to the maximum peak of the "modulation envelope."

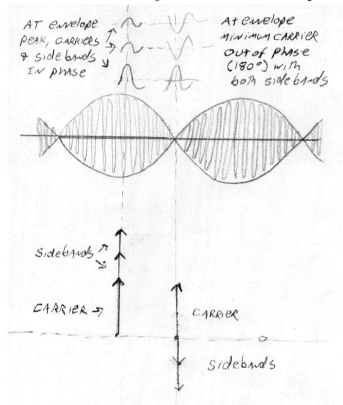

But remember that the frequencies of the three signals are not the same. Over time, over many cycles, these three signals will, because of the frequency differences, move in and out of phase. As they do this, the strength of the overall "envelope" will vary. Our reference moment described the envelope peak, but the three signals will also reach a point at which the two sidebands are both 180 degrees out of phase with the carrier; both sideband signals will be at their negative peaks when the carrier signal is at its positive peak. The carrier signal will, in effect, be cancelled out by the sidebands. This corresponds to the minimum point of the envelope. Then—because the three signals are of slightly different frequencies—they will all start moving back towards an in-phase condition. If you step back and look at the envelope, you will find—not surprisingly—that the frequency of the envelope variations is equal to the audio modulation frequency: 1 kHz.

I've noted the pleasant "quieting" provided by the carrier. But that pleasure comes at a big price. All of the voice information that you want to transmit is in the sidebands. There is a limit on how much power a given final amplifier signal can handle. If you are running AM, a significant portion of that limit will be devoted to transmitting the carrier. Even in the early days of "phone" amateurs realized that there was no real need for that carrier to be transmitted. And soon they found a way to get rid of it: If you use what is called a "balanced modulator" in the stage in which you mix the RF and AF signals, you will end up only with the sum and difference products of the mixing (more on this later). In other words, you'll end up only with the sidebands. Now you have what is called a Double Sideband Suppressed Carrier signal. In ham radio we simplify the name and acronym and call it Double Sideband or DSB.

Of course, the next step is to get rid of the other sideband. Doing so prevents power from being wasted in the final amplifiers, but even more important is the improvement in bandwidth. By getting rid of one of the sidebands, you cut your use of the precious electromagnetic spectrum by half.

There are a number of ways to get rid of one of the sidebands. A sufficiently narrow crystal filter is the simplest method, but those crystals cost money, and there is a cheaper but more complicated technique. I guess the boys at Hallicrafters were trying to make that HT-37 price competitive, because they went with the cheap but complicated "phasing" technique. There is no way they could have known how much technical head-scratching and teenage heartache that design decision would eventually cause. When my father would come home from work and find weird annotated electronic diagrams spread across the dining room table, a good number of them were undoubtedly block diagrams of phasing SSB rigs.

Here is the key idea behind the phasing technique: Look at the waveform diagram we have been discussing. Realize that in that waveform, there will be some point at which the two sidebands are out of phase with each other, some point at which the upper sideband will be hitting its positive peak just when the lower sideband is hitting its negative peak. We build TWO modulator circuits. We feed them with the same input signals. But we arrange it so that when the output of the first modulator is at our "reference moment" with both sidebands in phase, the output of the second modulator has sidebands out of phase with each other. When we combine (add) the outputs of the two modulators, one of the sidebands will be reinforced, the other will be nulled out. Because we are using balanced modulators, the carrier is also eliminated. DSB changes to SSB.

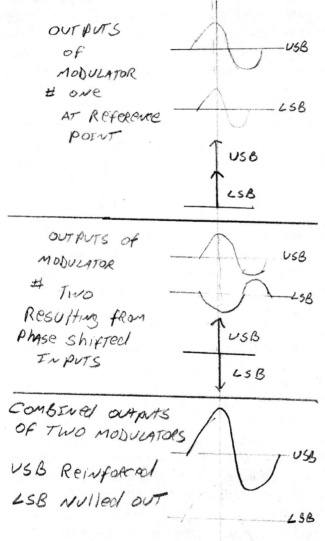

You get this very useful phasing arrangement by feeding the second modulator a slightly altered version of the RF and audio signals that are going into the first modulator. What is altered is the phase of the signals—at the second modulator the RF signal is shifted in phase by 90 degrees. And the audio going into that second modulator is also shifted by 90 degrees.

Great! SSB! But I didn't understand one key element of it: Try as I might, I just couldn't see how shifting the input signals by 90 degrees somehow led to the disappearance of one of the sidebands. It looked to me like you were just moving the whole thing 90 degrees. And a 90 degree shift won't null out anything, right? Oh how I agonized over this question!

The answer came years later, from a book that had been published a few years before. *Single Sideband for the Radio Amateur – A Digest of Authoritative Articles on Amateur Radio Single Sideband* was published by the ARRL in 1970. I probably picked it up at a hamfest in Virginia sometime in the late '90s. On the cover, an earnest young fellow spoke into a microphone. Above him was the oscilloscope waveform of an SSB signal, below him the photo of a homebrewed SSB transceiver. That fellow clearly had The Knack.

The first two articles in the book are by Donald E. Norgaard, W6VMH. Don's diagrams and text allowed me to finally understand my beloved HT-37.

It turns out that when a mixer spits out its sum and difference frequencies, they have a specific phase relationship with the input signals. The reason for this can be found in the mixer math, in the work of Fourier, in the trig. If you shift the RF signal going to that second modulator by 90 degrees and also shift the audio input by 90 degrees (both relative to the signals going into the first modulator), you end up with sideband outputs that will have a very convenient phase relationship to the sideband outputs of the first modulator. When you combine the outputs of the two modulators one sideband will be reinforced, and one will be nulled out.

Here's the key point I was missing: When you shift the two inputs by 90 degrees, you are NOT simply shifting the entire output signal by 90 degrees. Because of the relationship between the phase of the input signals and the sideband products, you end up with sidebands that have a phase relationship to the outputs of the first modulator that allows for the nulling out of one of the sidebands.

That is not very intuitive. It is not obvious (at least not to me) that by shifting inputs 90 degrees, you will end up with the kind of outputs that will reinforce one sideband while annihilating the other. I wonder what the history of this circuit is; I wonder if it resulted from cut-and-try experimentation, or if it popped out of the trigonometric formulas.

Whatever the origin of the circuit, with the insights provided by Don Norgaard, I could use my HT-37 without feeling like a complete appliance operator. Thanks Don.

****

While AM was once the king of the amateur phone bands, by the time I got on the air in the 1970s, SSB ruled the roost and AM operators were in minority status. Well, truth-be-told, sometimes it seemed more like cult status. Single sideband operators often spoke disparagingly of the aficionados of "Ancient Modulation" and complained that they used too much of the band. Looking down their technical noses at the inferior audio quality of the SSB rigs, the AM operators referred to the rival mode as "Single Slopbucket." It all got kind of silly.

There is a lot of debate about this, but I think the audio quality of an AM signal is better than that of SSB. The SSB rigs often have more constrained audio filtering than the old AM rigs. With an AM signal, if the receiver is slightly off frequency, it will be barely noticeable, but be a bit off frequency with an SSB signal, and your interlocutor sounds like Donald Duck. When you are receiving an SSB signal, you make use of a carrier that is generated in your receiver—in an AM signal, the carrier and the audio sidebands have exactly the correct phase relationship. Not so with SSB, and this may result in some additional distortion. And there is that very pleasing "ker-chunk" sound followed by the silence induced by the carrier.

There are endless discussions about the relative technical merits of SSB and AM, but the real reasons that guys stick with the old technology are social and emotional: We like the old rigs, we like the sound of AM, and we like talking to guys who share our enthusiasms.

There are some real characters on the AM frequencies. Many of them are the same guys who I'd listened to 25 years earlier as a teenager. After being away from the AM frequencies for many years, when I did return I immediately recognized some of the call signs, and a few of the voices. They'd been talking to each other for so long that some of them had even slipped into a very distinctive AM argot.

The level of technical knowledge and expertise among the AM operators was usually higher than that of your average ham. The older rigs required maintenance, and most guys are engaged in a never-ending campaign of modification and improvement, especially in areas that would result in improvements in audio quality. As with the homebrew radio, AM is not plug-and-play. When you make contact with someone and he tells you that he is running a modified DX-100 and a Hammarlund Super Pro 600, you know that this guy knows one end of a soldering iron from the other. (What a great name for a rig! Shep said that "Super-Pro" was "a name with gonads.") This gives AM operation a strong feeling of fellowship and fraternity.

Within the fraternity, there is a hierarchy. A hierarchy of rigs. At the top there are the big signal guys. These fellows are running very powerful, plate-modulated AM rigs. We are talking here about DX-100s, Viking Valiants, military surplus T-368s. There is even a monster rig called the Desk Kilowatt: 1000 watts of RF power built—you guessed it—into a stylish (in '53) desk.

My HT-37 generated a fairly good AM signal, but it wasn't really an AM rig. It was really an SSB rig that could be made to generate AM. I wanted to be more a part of the AM mainstream, so I started to put a real AM station together. I took the best parts of the two hamfest DX-60s and combined them into one working transmitter—this was paired up with my re-born green-eyed VF-1. For the receiver, I started out using the Drake 2-B, but this didn't really work out. It pains me to say something bad about my 2-B, so I will turn my criticism into a pseudo-compliment: The Drake 2-B is so selective, that even at its widest setting (3.6 kHz) its bandwidth is still a bit constrained for double-wide AM signals.

Luckily I had an old Hammarlund HQ-100 receiver that was, as the Old Time hams used to say, "as wide as a barn door." This Hammarlund was an economy receiver. There were no expensive crystal filters to be found in that device. AM Shortwave broadcast stations and amateur AM stations all sounded great, but unfortunately sometimes they all sounded great at the same time, at the same spot on the dial.

Perhaps it was my need for improved AM reception that got me involved in another receiver project...

This is the story of a receiver that came together through a process very similar to spontaneous combustion. In the corner of my Virginia basement hamshack, I'd been slowly but surely filling up a cardboard box labeled "good junk." (I'm sure many readers have similar boxes.) In the aftermath of each hamfest, a few new (old) pieces would be added to the pile. Variable capacitors, tubes, sockets, transformers - you know, the good stuff. Close-by, a collection of ham radio magazines from the '50s and '60s grew at a similar rate.

The accumulation seemed very innocent. Little did I know I was approaching critical mass.

The spark came from the April 1966 issue of *QST*, page 49. In his "Beginner and Novice" article, Lew McCoy presented a little three tube superhet receiver for 80 and 40 that promised to be "a real performer." This was the receiver companion to the "Mighty Midget" transmitter project that had been presented a few months earlier. (The transmitter was presented as part of a contest: The first novice to finish building it and make contact with stations in ten ARRL regions won!) I liked the looks of the circuit, but at first I was uneasy about the metal work that would be required. I have some deep, primordial ham radio memories of hacking away with inadequate tools at the chassis of that homebrew power supply for the HW-32A. The experience had made me a believer in PC boards. But this receiver circuit was very enticing. The thought of tuning in CW, SSB, and AM signals through all of 80 and 40 on a homebrew rig was very appealing. I could even do some short-wave listening on 40. But still, there was the metalwork problem...

Suddenly, while pondering a photograph of Lew's receiver, it occurred to me that the size and layout was remarkably similar to that of an old Heathkit rig that I'd picked up at a hamfest. This little portable transceiver was part of what came to be known as the Benton Harbor lunchbox series. This one was for the six meter (50 Mhz) band and was called (you guessed it): "The Sixer." The wheels started turning. A quick check of the junkbox revealed that I had most of the parts on-hand. A look at the Sixer chassis showed that it would save me almost all of the hated metal work - I could make use of holes that were cut in Benton Harbor, Michigan during the 1960s. Soon I knew for sure that that Sixer would never again vibrate the ether at 50 MHz. I justified my decision by noting that the rig had already been modified almost beyond recognition and would not therefore be of historical value (I swear). I started gutting it with gusto.

The old handbooks recommend covering a new chassis in wrapping paper so that the placement of every socket and screw can be carefully planned. (I often wondered if there were real hams who were patient enough to actually follow this good advice.) On this project I just took the major parts and moved them around on the chassis until I had a satisfactory layout. I tried to keep the tubes away from the local oscillator coil and capacitor. There was one big hole where the old electrolytic capacitor had been; I used some spare PC board material to patch over this hole, using it as the mount for the FT-243 crystal holders for the IF filter. Underneath, I just threw in some terminal strips wherever I thought they'd be useful (those with lots of grounded terminals are the best).

My junk box provided most of the components, and a conveniently-timed Manassas, Va. hamfest helped me fill some of the gaps in my parts list. The slug-tuned coil for the BFO was a particularly happy discovery at that 'fest. I used a few more of the variable caps out of Pericles' old Swan 240 transceiver. The power and audio transformers came from the Sixer.

I had to make a trip to the local pharmacy to pick up the empty pill bottles for the coils. I figured they would be suspicious about my request for what could, I suppose, be considered drug paraphernalia, so I brought with me a copy of the April, 1966 *QST* and showed the pharmacist the picture of the Mighty Midget chassis, with its pill bottle coil forms. I immediately "scored" four free pill bottles! It was a lot of fun to wind the coils, and I think they look very fine on the chassis. They add some homebrew panache to the project.

The really difficult-to-find components were the two 455 kHz crystals for the lattice filter. I discovered that these rocks have become quite rare. A major manufacturer of crystals (who will remain anonymous) gleefully offered to make me the rocks - for a mere $75.00 dollars each. Ouch! But then the fraternity of cyber-space solder melters came to the rescue: James, W5LWU, donated three surplus FT-241A crystals. Unfortunately I was unable to get them to function in the filter. I get the impression that these rocks do not age very well. I'm also told that even under the best of circumstances homebrew lattice filters were hit and miss affairs, with more misses than hits.

Anxious to get the radio going, I substituted a 455 kHz IF transformer for the crystal filter. This was very easy, because the IF transformer fit perfectly into the two adjacent FT-243 crystal holders on the chassis. A quick rearrangement of a few leads and I was in business. This setup left the receiver a bit broad, but I kind of like it that way. Phone signals (particularly AM sigs) sound very nice. While I couldn't get the complete "single signal" effect with this system, with careful placement of the BFO frequency I did notice a very significant attenuation of the audio image.

I'd originally planned on making a completely new front panel, but I came up with an alternative that allowed me to make use of the old Sixer panel: Home Depot sells some very light, thin aluminum sheeting material (used for storm or screen doors). Using tin shears, I cut out a "false front" for the Heath front panel. I got mechanical stability from the sturdy Lunchbox panel while covering up the scars of modifications from days-gone-by. I secured the main tuning cap to both the chassis and the front panel - this provided a very noticeable improvement in stability. I found that I really didn't need a reduction drive for the main tuning cap. I put a DX-60 knob with a pointer on it. An old CD cut in half and affixed to the front panel serves as the dial.

The receiver went together very smoothly. Debugging was unusually easy. I was amazed to find absolutely no unwanted oscillations to be stamped out - this was particularly surprising given the small size of the chassis, the use of pill bottle coils (versus toroids), and the fact that I didn't pay a lot of attention to shielding. I had to experiment a bit with the coils for the local oscillator to get it vibrating on the proper frequencies. I also had to put a little trimmer cap on the tuned circuit in the RF amplifier's plate to get it to track well with the grid circuit. Putting the RX inside the cabinet seriously detuned the BFO, so I had to come up with some way of adjusting (from outside the cabinet) the chassis-mounted slug-tuned inductor. I ended up putting another hole in the already well-perforated cabinet. With a slightly widened plastic nut-starter I could easily reach down and adjust the BFO. Coming up with solutions like that is one of the joys of homebrewing.

I added a few modifications to Lew's original design. With AM operation in mind, a "BFO OFF" switch was obviously a necessity. I also put a fuse in the power supply circuit. I muted the receiver (during transmit periods) by lifting off ground both the IF amplifier's cathode and the ground connection on the RF gain control.

I thought about adding an additional stage of AF amplification but things were getting a bit cramped so I decided to leave well-enough alone. When I want to use a loudspeaker I simply plug one of those little computer speakers with an internal AF amp into the headphone jack. I realize many tube purists will find this distasteful, but let me point out that I briefly considered an even more unpleasant option: placing a little LM386 AF amplifier chip in there among the 6U8s, variable caps and pillbox coils. Somehow it just didn't seem right - the external AF amp option seemed to be the lesser of two evils.

Lew McCoy was right when he promised that this receiver would be a real performer. The two tuned circuits in the RF amplifier seem to take care of all the image problems (because mixers produce sum AND difference frequencies, unless there is some filtering before the mixer the receiver will respond to signals at two different frequencies). They also very effectively kept the intense Northern Virginia AM broadcast energy out. Sensitivity was very good - I often had to back off on the RF gain and I could easily copy the Australians on 40 SSB in the morning. After a warm-up period the receiver always became very stable.

One of the best parts of this project was the support and encouragement that I got over the internet from the worldwide fraternity of solder melters. Of course, this internet link is something that didn't exist when Lew wrote his article in 1966, or when I was struggling with my ill-fated receiver project in 1974. We used the rec.radio.amateur.homebrew USENET group and the GLOWBUGS and HOMEBREW mailing lists. It was great fun to share progress reports with a large group of fellow radio fiends. As a result of this Internet chatter, a number of other Mighty Midget receiver projects were started: Sandy, W5TVW, experimented with 6T9s in place of the 6U8s. Rod, N5HV, was building one on the chassis of a dead Heath Twoer (will there be no end to this carnage?). Giovanni,

IT9XXS, planned an Italian version. Jose, EB5GAV, gathered 6U8s in Valencia, Spain. Cedrick, N9YXA, used a Mighty Midget receiver project to take his mind off the snows of winter. Collin, N4UTA and Chris, KX0Y were both planning MM projects.

Michael Hopkins, AB5L, found a Frank Jones article that suggested a way of powering this receiver without a transformer: He used rectified and filtered AC line voltage to the plate with the AC voltage dropped to filament levels via a light bulb in series. (You get a desk lamp as part of the deal!) Finally, Eddy, VE3CUI, was delighted to learn that he is not alone in his enthusiasm for this simple receiver. Stimulated by the 1969 ARRL Handbook, Eddy had built his version of the MM three years before I built mine. His is a very "souped up" version that operates on 160 and 80 with additional stages of IF and AF amplification.

This was my first tube-type construction project and I really had a lot of fun with it. When you build solid state gear on PC boards, it's all very one dimensional, very flat. But this kind of tube project is very 3-D. As you add components above and below the chassis, you really get the sensation that you are building something substantial. And of course there are other aesthetic rewards: the warm glow of the firebottles and that wonderful smell that comes from oil and rosin heated by filaments.

Soon I had the homebrew Mighty Midget receiver paired up with my hamfest DX-60 and the born-again VF-1. While I wasn't at the "big signal" level, I felt like I was operating on AM with style. As is so often the case with ham radio, the medium became the message: when I talked to other hams, much of the conversation was about the gear that we were using. And my setup gave me a lot to talk about. People loved hearing about the homebrew receiver, the salvaged transmitter, and the re-born VFO. But beyond just giving us something to talk about, using this kind of gear helped establish and solidify a sense of fraternity and belonging among the people who met on the AM frequencies. We'd all done something more than just handing over our credit cards to get on the air. Our paths to the airways had included struggles with metal chassis, hot tubes and molten solder. We'd all participated directly in the construction or restoration of the machines that produced the waves that carried our melodious voices into the homes of our un-met friends.

As much as I liked the old firebottles, around the time that the Mighty Midget receiver project was finished something wonderful happened that caused me to shift away from high voltage circuitry. My son Billy was born on November 5, 1997. Suddenly, I felt much more necessary and needed in this world. The risks associated with working on high voltage equipment started to seem a bit unreasonable. I was still on my early-morning-in-the-shack schedule, but now, like all new dads, I was often sleep-deprived. Fatigue does not mix well with high voltage. And looking ahead, I knew that before too long, Billy would be with me at the workbench, watching me work on the gear. Having circuits with 800 to 1000 volts DC floating around didn't seem consistent with the massive child-proofing campaign that I was waging elsewhere in our house. So I decided to make the Mighty Midget my last tube-type piece of gear. Fatherhood had moved me into the transistor age, into the safer world of 12 volt power supplies and solid state components. This move didn't bother me at all. It was time to make the switch, time to jump from the 1950's into the modern age (or at least something closer to it.)

Billy did indeed take an early interest in my hobby activities. When he was still an infant, we discovered that one way to calm him was for me to step out to the driveway with him at night. I'd sit on the hood of the car with him in my arms and we'd look at the moon and the stars. He also showed very early "car-guy" tendencies: He'd put himself at the window and keep track of the vehicles passing in front of the house. One of his first words was his version of car ("kaitje"). Soon he was down in the shack with me, happily watching as I played with the electronics.

My switch from tubes to transistors was encouraged by the emergence at about this time of a new radio club that was almost completely solid state in its orientation: NOVA-QRP. The Northern Virginia QRP Club was sort of an un-club. The organizers decided against organization. There would be no dues, no formal membership lists, no officers, no formal meetings, no minutes, and definitely no Roberts Rules of Order. For someone like me who spent much of his working day dealing with meetings and bureaucratic struggles, NOVA-QRP's approach was very appealing. The last thing I needed at the end of the day was another meeting. NOVA-QRP would simply get together once a month or so at a restaurant ("Mama's") and talk about our low power radio projects. Guys would bring in gear they were working on, parts would be traded and donated, and all would benefit from prodigious quantities of good fellowship.

We'd gotten quite settled in Northern Virginia, but by early 2000 the Foreign Service storm clouds were gathering again—I knew that during that summer the tornado would hit again, lifting us up, wiping out the shack and wiping from that small house in Northern Virginia all evidence of our having been there.

This one was going to be an especially anxious move for us. Billy was only two years-old, and my wife was carrying soon-to-be born Maria. I quickly discovered that the Foreign Service assignment process is a lot more stressful when young children are involved. Possible assignments to several different places popped up out of the bureaucracy, but then quickly faded away. Monterrey, Mexico was, for a time, on our horizon. But that fell through. Madrid looked good for a while, but someone else got that one. A number of exotic third word locations were proposed, but during visits to the relevant web sites we could practically hear the gunfire in the background—they didn't seem very "child friendly." Then, at a State Department Christmas party, someone mentioned that there was an opening that might be good for us.

CHAPTER 5
MID-ATLANTIC OUTPOST
AMATEUR RADIO FROM THE AZORES

Sometime about a million and a half years ago, some forgotten genius of the hominid world did an unexpected thing. He (or very possibly she) took one stone and carefully used it to shape another. The result was a simple teardrop-shaped hand axe, but it was the world's first piece of advanced technology. It was so superior to existing tools that soon others were following the inventor's lead and making hand axes of their own. Eventually whole societies existed that seemed to do nothing else. "They made them in their thousands," says Ian Tattersal... "It's strange because they are quite intensive objects to make. It was as if they made them for the sheer pleasure of it. ...The axes became known as Acheulean tools.... These early Homo sapiens loved their Acheulean tools... They carried them vast distances. Sometimes they even took unshaped rocks with them to make into tools later on. They were, in a word, devoted to the technology."

—From "A Short History of Nearly Everything" by Bill Bryson.

"Why build a receiver? Why do you want to build it?" asked Harish, an old friend, when he spotted us struggling over the DC40 one evening. I didn't have an answer to this question and considering the amount of work piled this quarter, it appeared to be a sensible thing to ask. I think this question is answered by us all in different ways. My personal answer would be because we human beings are fundamentally tool builders. We have an opposable thumb that allows us to grip the soldering iron.

—Ashhar Farhan, Indian Radio Amateur, Engineer, and designer of the BITX 20 transceiver

Looking back on it, I guess my amateur radio background probably had a sub-conscious impact on my decision. For most normal people the words "volcanic island in the middle of the North Atlantic" probably would raise some concerns. But for me, I found the coordinates and the geography somehow alluring. I wasn't consciously thinking about propagation patterns, or the opportunity to live in a place that seemed to come straight out of the pages of "How's DX" ("And here's Bill Meara, N2CQR, at his shack atop volcanic Sao Miguel island in the Azores chain..."), but truth-be-told, ham radio probably was one of the factors that caused us to move to the Azores in summer of 2000.

Maria had been born in March, a few days after our assignment was made official. I joked with the staff in the delivery ward that she'd been waiting to see where we were going before formally signing on for the trip. Billy was very focused on the weather of our new home, and as we prepared for the journey he frequently chimed in with "No snow in Ponta Delgada!"

The house of the American Consul in the Azores was located in the middle of Ponta Delgada, a small town on the south coast of Sao Miguel island, the most populated of the nine islands in the Azores chain. The island is more or less rectangular, stretching east-west. It is about thirty miles long and ten miles wide. The island is essentially the conjoined tips of three volcanic peaks that rise dramatically from the ocean depths. There is a semi-dormant volcano to the Northwest that is visible from the town. There is another large volcano to the Northeast. In Ponta Delgada you can never forget that you are on a small island—the ocean is almost always in your field of view to the south, and you know that a very short drive over the hills will bring you the Northern version of that same panorama.

Almost everything on the islands is made of or from volcanic material. Walk up an exterior staircase and you will notice that the masonry is pockmarked—the little holes are the remnants of volcanic bubbling. At the beaches, you can see where the lava flows hit the water and suddenly cooled into rock (Billy loved climbing on those rocks). The sand is black. Through Billy we discovered that it is quite metallic; one day the little guy brought a toy magnet with him to the beach. "Look daddy, it is picking up the sand." Of course, I immediately thought of the positive impact that this highly conductive soil would have on radio propagation.

The house was absurdly large, much too big for us, big way beyond the possibility of being cozy, too big to heat properly. But it came with the job and we were stuck with it. There were many rooms in the place that could have been pressed into service as the radio shack. I could have used one of the big upstairs bedrooms, or one of the guest rooms. But somehow I gravitated toward a really ugly storage room below the kitchen, off the garage. Somehow it seemed right. It looked like the kind of room that wouldn't be bothered by flying blobs of molten solder.

Our Shack in the Azores – Operating position

The packing crates arrived and the new shack began to fill up with the old familiar radio gear. For most radio amateurs the most important part of the shack is the operating position—the place where they have the radios set up for on-the-air conversations. But among homebrewers, the operating position has a competitor: the workbench, the place where solder is melted and circuits are created. My workbench was a Father's Day gift from my wife. It is a Home Depot product of exactly the right dimensions. Six foot long, four foot deep and a few inches higher than a standard table, this wonderful wooden bench became the cradle for my electronic creations. It took the place of honor in the new shack.

The Consulate acquired the necessary forms and soon I had an Azorean call sign: CU2JL.

When my thoughts turned to antennas, I had to face up to a rather special and unfamiliar problem: prominence in the community. All my life I had gone out into the yard and had spent hours throwing rocks with strings attached into the branches of trees. This rock and string throwing was often accompanied by much cussing and tangling, and occasionally resulted in minor injuries. This was clearly the kind of activity that would cause passers-by to lament the cut-backs in the budgets for our mental health facilities.

I'd never before been in a position where anybody really cared whether or not I was nuts, but in the Azores my anonymity was gone, and I had to pay some attention to public perceptions. I had to be a bit more discrete about my antenna work, so I started out with a simple dipole. I had larger ambitions, of course, but I figured I'd start out slow and let the locals gradually get used to my electromagnetic eccentricities.

During those first few months in the Azores there wasn't a lot of time for ham radio. We were busy getting settled in the new place, finding a pre-school for Billy, getting settled in the new job, figuring out how to run the mansion, etc. But in this entirely new place, amateur radio did, during those early months, provide that sense of continuity and connection with the old and familiar that has always helped me adjust to a new environment.

And many adjustments were required. We got Billy enrolled in a local nursery school, but he was the only kid there who didn't speak Portuguese. No one there spoke English. The language gap was bridged by the loving care of the teachers (hugs require no translation). In those early weeks I consoled Billy about his language difficulties. I told him that he should not feel bad about his inability to speak to the other kids. I explained that he just hadn't learned Portuguese yet. "No dad, that's not why I don't talk in school. I don't talk to the other kids because my teeth are too big." That was his way of understanding the language barrier.

Maria was too young for pre-school, so her introduction to Azorean life came from Margarida and Paula, the Azorean women who worked in the Consul's residence. Maria would sit with them as they did their daily chores, and it was inevitable that she would pick up Portuguese from them. One of Maria's first words was "'ta?" the interrogative that Margarida and Paula used when answering the phone. By the time we left the island both Billy and Maria were very comfortable with Portuguese—Billy had picked up the peculiar accent of the island (Micalense)—it really freaked people out when this little American kid spoke in their island accent. And he understood the accent just as well: One day as we bought supplies for a fishing trip, the very Azorean bait and tackle guy warned us in rapid-fire Micalense that we should be careful with any red-colored fish—they have poisonous barbs. I guess Billy had some doubts about my language skills, because as soon as we got into the car he gave me an emphatic and verbatim translation of the shopkeeper's warning.

When there finally was time for radio, the first thing I set up was, of course, the Drake 2-B. One of the local shops carried ham radio magazines, and from them I learned that on November 16, 2000 the new Phase 3-D amateur satellite had been launched from the European Space Agency site in Kourou, French Guyana. This was big news for amateur satellite fans. Unlike the low-orbit RS-12 and RS-10 satellites, P3D was going into a high "Moliyna" orbit. Not quite the high geo-stationary orbit that allows communication satellites to match the earth's rotation and hover permanently above one region, the Molinya orbit had most of the advantages of the high orbit (long time periods of availability, the possibility of very long distance contacts), but with the additional advantage of allowing amateurs around the world to share the satellite. P3D had an elliptical path with the earth at one end of the ellipse. It would shoot up to a point some 30,000 miles above the earth, then drop down to within 3,000 miles of the surface before shooting up again. It would take a long time—many hours—to rise and fall, and during these rise and fall periods it would appear from Earth to be at a fixed point in the sky. But on its next rise and fall cycle, the earth would have rotated below it, and another group of amateurs would be having their turn.

P3D was not the first high earth orbit amateur satellite. AO-10 had gone up in 1983, but its main computer had failed in 1986 and the satellite was only sporadically available for use. AO-13 had been launched in 1988, but there were problems with its orbital path and it burned up in the earth's atmosphere in 1996. While in the Dominican Republic I'd used my Ray Gun antenna to listen to AO-13 and AO-10. I'd tried without success to build a transmit system that would have allowed me to make contacts through AO-13— the kind of Ultra High Frequency gear required for this kind of operation was probably beyond my technical capabilities.

P3D promised to open a new era for satellite enthusiasts. It would be a powerful satellite, and it would allow operations on many different bands of frequencies. Enthused, I quickly rigged up a frequency down-converter that would allow me to listen to the telemetry beacon using my trusty 2-B. I started hearing the telemetry from P3D on December 3, 2000. It was a lot of fun to sit in my new shack on that small island and listen to signals coming in from an amateur-built device that was 15,000 miles or more out in space. Later that first morning, while trying to tune in the telemetry, I suddenly heard the voices of radio amateurs. This was strange, because the P3D satellite had not yet been made available for amateur conversations. Then I realized that I was hearing signals from *another* amateur satellite: Low earth orbit RS-13 was passing overhead, and it was beaming to earth signals from the 15 meter amateur band.

Monitoring the telemetry from the new satellite became part of my daily routine during that first winter in the Azores. But one morning, I suspected that something was wrong. The telemetry signal just wasn't there. I checked the computer for the satellite's position, and confirmed that it was overhead, but the familiar warbling sound was not coming out of the 2-B. A few days later, word came over the internet that something had gone horribly wrong when ground controllers had tried to use the on-board thruster rockets. Later, we found out that problems with a valve had caused an explosion that had greatly diminished the satellite's capabilities. Of course, that malfunction marked a great loss to hams around the world, and must have been heartbreaking to those volunteer amateurs who had worked so hard to put that spacecraft together. But the fact that I—sitting off by myself in my remote island ham shack—had independently and directly detected that something was wrong on that spacecraft was, in a bittersweet way, kind of satisfying.

In those early months in the Azores, I also started making contact with the local radio amateurs. As in the Dominican Republic, and probably for very similar reasons, amateur radio was popular in the Azores. As we drove around town, I spotted the antennas that were the tell-tale signs that I was not the only radio fiend on Sao Miguel. On the road up to the supermarket, there was a small building that was remarkably similar to the little radio shack back in New York that had helped put me on the path to radio addiction. I stopped by one morning and established contact with the local hams.

A few days later I had a visitor at the Consulate. Our Portuguese employees told me that he was a retired gentleman who shared my interest in radio. This was my first meeting with Messias Moniz, CU2BJ.

Messias was in his late 60's; "wiry" described both his avocation and his build. He had the rugged, leathery features of someone who'd spent a lot of time in sea-blown winds. Messias was just stopping by to say hello, to welcome a fellow amateur to the island. He brought with him a collection of ham radio magazines and literature about the club. As is usually the case with first meetings like this, I didn't realize at the time that Messias would become a good friend.

During those first few months in the Azores I was on the air with an old Heathkit HW-101. After a nostalgia-driven spree in which I'd bought up much of the gear that I'd used as a kid, a second, more insidious and expensive phase of the Boatanchors disease had kicked in. In this phase I'd started buying not just the gear that I'd actually used in my electromagnetic youth, but gear that I'd lusted for but could not then afford. In my case the best example is the Heathkit HW-101 transceiver. I always thought that was a beautiful piece of equipment. At our Crystal Radio Club Field Day expeditions to the Nike-Hercules Missile site, the club had set up one of these rigs as the novice station. It looked quite a bit like the HW-32A for which I'd risked electrocution in the power supply project, but unlike the HW-32A this 101 covered all of the HF ham bands. I'd wanted one. And now I could afford one. I got mine through eBay.

The HW-101 was designed for economy and priced to sell. It quickly became the most widely-produced high frequency transceiver in history. But the cost cutting had its down side: this rig didn't age well. The plastic in it got brittle. Many of the resistors probably should have been of a higher wattage. Over time lots of problems cropped up, and, for me, the HW-101 was soon in the "high maintenance" category.

Early-on, problems with the Variable Frequency Oscillator became apparent. This was the main tuning circuit for the transceiver. Instead of smoothly tuning across the frequency band, the oscillator seemed to jump and skip erratically from frequency to frequency. This was no good at all—something had to be done. I started studying the schematic and was contemplating major surgery when I decided to first consult with the on-line community of old-radio fans.

I posted a message describing my problem to one of the mailing lists, and quickly an answer came back from somewhere in the United States. An old-timer well familiar with the rig told me to put my soldering iron away and to instead apply a few drops of oil to the reduction drive that connected the main tuning capacitor with the front panel tuning knob. That was all that was needed. It always pays to consult with the experts.

The brittle plastic problem soon reared its ugly head. The HW-101 was built before the age of glowing numerical frequency displays. Behind the front panel, just above the main tuning knob, a small disk with the frequencies printed on it appeared through a small window in the front panel. This disk turned as you tuned the receiver. To allow the operator to calibrate the dial (to make sure that the frequency in the little window corresponded with the actual frequency to which the transceiver was tuned) Heathkit had a little mechanical device near the main tuning dial that allowed you to hold the "displayed" frequency in one place while you brought the actual frequency of the transceiver (of the VFO) into alignment. This was called "the dial clutch." It was a second small plastic disk; this one was moved via a plastic button that extended through the front panel. One day I pushed that plastic button and heard a disheartening crack from inside the rig. This was not good. On inspection I found that my clutch had shattered. I soon found some plastic of similar consistency and fashioned a replacement clutch. This worked fine, but I started to have my doubts about the HW-101—this was a radio transceiver in which I'd had to change both the oil and the clutch. Sometimes it seemed as if I was working on a problematic old Chevy. I started to get tired of all the HW-101 geriatric care.

Besides, it was time for a change. I was trying to transition away from tube-type gear like the HW-101. And I wanted to build something completely new. I felt the urge to melt solder, the urge to engage in some Mid-Atlantic homebrewing.

You might suspect that a small island in the middle of the Atlantic Ocean would not be the best place to build radios. There were no RadioShack branch stores. There were some very enthusiastic hams on the island, but most of them were not homebrewers, so you wouldn't have many locals to turn to if your project ran into trouble. One day while home for lunch, three year-old Billy came to me with a question. He'd been watching American cartoons via our Armed Forces Radio and Television Service satellite link, and he'd heard a phrase that he didn't quite understand. "Dad," he asked, "Where is the middle of nowhere?" I couldn't resist. "Here son," I said pointing down, "It's right here!"

But, of course, by 2001 the internet had taken care of this "middle of nowhere" problem. All the parts I needed were available via on-line vendors. And overlapping networks of mailing lists and USENET groups meant that advice, information, and words of encouragement were all just a few keystrokes away.

I knew that in an isolated place like the Azores, the internet would play a key role in any homebrew projects. I'd set up my first website while back in Virginia. I'd had great fun using the site to share my radio adventures with fellow geeks. For example, with the help of my friend Tyler Dunn, we had digitized the recordings of my conversations with the astronauts—when placed on the web site, these clips had drawn listeners from around the world.

Blogging is now all the rage, but I guess it is fair for me to say that I was blogging before the term had been invented. In April 2001, when I started my first Azorean construction project, I sent progress reports to the QRP-L mailing list, an e-mail linked group with about 700 members worldwide. I would also put these progress reports on my web page. That may have been one of the first ham radio blogs.

If you've made it this far in the book, you probably will be interested in my Web-based, blow-by-blow description of what it was like to homebrew a piece of amateur radio gear in the Azores islands. Here it is. I've translated the otherwise incomprehensible ham slang and acronyms, and I've inserted a few photos and diagrams.

April 24, 2001:

I feel the urge to melt solder. And I want to build a phone (voice) rig. I've always wanted to get on 17 meters (no contests!). I can listen to that band on my Drake 2-B. Here's my plan:

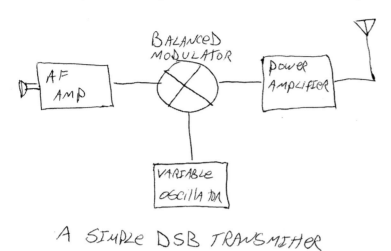

A SIMPLE DSB TRANSMITTER

Solid State double-sideband (DSB) suppressed carrier transmitter (TX) based on Doug DeMaw's circuits. Very simple. Sort of a direct conversion transmitter. The oscillator will be on the transmit frequency. I think I'll start out with a variable crystal oscillator (VXO) (I have the rock). Solid state power amplifier (PA). (Another option I considered would be to put the boards on an old DX-60 chassis and rebuild the 6146 tube power amplifier stage... but I think that might be overkill.)

I'm going to start out just building a transmitter. I'll build the oscillator first, then the DSB generator. Finally (!) the PA. I'm going to leave room in the cabinet for receive circuitry. It would be very easy (and sort of appropriate) to build a matching direct conversion receiver, but I've had bad luck with them and may decide to build a real superhet receiver and make the final rig a Transmitter - Receiver (not a transceiver).

I have an old Bud cabinet that's about the right size. I'm thinking of using the Manhattan technique on the circuit boards.

What do you guys think? Anyone out there built this kind of rig?

73 from the Azores. Bill CU2JL N2CQR

April 28, 2001

I stopped off at a hardware store yesterday and bought some super glue. This morning I used it to build the first stage of the 17 meter DSB rig: the Variable Crystal Oscillator. I used the circuit from Doug DeMaw's "Barebones superhet" (June 82 *QST*) Very simple. I used an MPF102 field effect transistor instead of the 40673 dual gate MOSFET. The stage went together very quickly - Manhattan style is the way to go! [It is called "Manhattan" because when you look at the board from above, the square and rectangular shapes of the isolation pads look like the street pattern of Manhattan Island.] I'd been listening to 17 meter SSB as I worked. It was very satisfying when, at the end of the morning, I connected a 12 volt battery to the circuit board and heard my new oscillator through the receiver. I got very good "swing" in the crystal—looks like > 20 kHz.

I think I'll build a direct conversion RX to go with this rig—this way I'll be listening to both sidebands at the same time and will be better able to avoid interfering with nearby stations.

73 from the Azores Bill CU2JL N2CQR

May 5, 2001

There has been steady progress on my 17 meter DSB rig.

With the VXO completed (or so I thought!), I moved on to the microphone amp and the balanced modulator. The mic amp was a 741 op amp—construction and testing were uneventful. I used Manhattan style construction—the chip was crazy glued to the board, leads up.

The balanced modulator went together very well. Given that I'm 900 miles from the nearest electronics store, I was very pleased to find that I had all the needed components in my junk box. I used hot carrier diodes that a very kind member of the Northern Virginia QRP club had been giving away at a club luncheon. I picked two that had forward resistances that were very close. I didn't have the 250 ohm balancing potentiometer (variable resistor) that Doug DeMaw's circuit called for, so I used a small 2K pot with an appropriate fixed-value resistor across it. I tried to keep the physical layout of the balanced modulator symmetrical.

A problem cropped up when I tried to test it. When I connected the VXO to the Local Oscillator, the VXO shut down. I played around with some of the feedback circuit values, but no joy. So I just used some of the extra space on the VXO board to build a simple JFET amplifier circuit (really just for isolation). This solved the problem.

I could now test the balanced modulator. Putting an RF probe connected to a digital multi-meter at the modulator's output, I turned on the VXO and tried to see if I could balance out the carrier. Worked like a charm! Just about in the center of the pot's movement, output dropped to near zero. Then, to see if an incoming audio signal would UN-balance it, I injected some noise into the "mic in" port. Output jumped considerably.

Next I built the first RF amp stage, just a simple Class A amp using a 3904 transistor. I goofed by getting the collector and emitter confused (it didn't amplify too well that way!) but once I got it straightened out, I was getting the desired gain.

Then came the cool part of the morning: I connected a microphone to the modulator and a six foot piece of wire at the output of the RF AMP. I tuned my Drake 2-B to the output frequency. Then I balanced out the carrier. Then I spoke into the mic. DSB through the Drake 2-B! I could hear some RF getting into the input, but other than that it sounded pretty good. Some shielding will take care of the RF problem.

NEXT STEP: Power amplifier! I'm looking for a circuit that doesn't require any exotic parts (defined as parts not currently in my junk box—some of those toroidal cores used by W1FB are just not on hand). Any suggestions? Also, I'm short on 1 mH RF chokes. Any thoughts on how I can wind my own?

May 8, 2001

I noticed that the voltage level at the local oscillator port on my balanced modulator was very low. Doug DeMaw wrote that it should be 1 volt rms, but I'm measuring only .1 volt.

I think the problem is that my VXO has a high impedance output while the LO port is low impedance. The VXO has a JFET amp and was designed to connect to the gate of a MOSFET. The LO port is one coil on a 12 turn trifilar balanced modulator transformer.

With no load on the VXO output I get about 1.89 volts rms. The problem, of course, arises when I connect the output to the balanced modulator LO port - then I only get around .1 volt at the port.

I've been playing around with an impedance matching transformer using a toroid core. With 12 turns on the primary and 3 on the secondary I can get the input voltage up to about .5 volts.

Am I heading in the right direction? Any other suggestions on how to handle this? Could it be that I just need another stage of amplification in the VXO, or should I continue to experiment with the transformer?

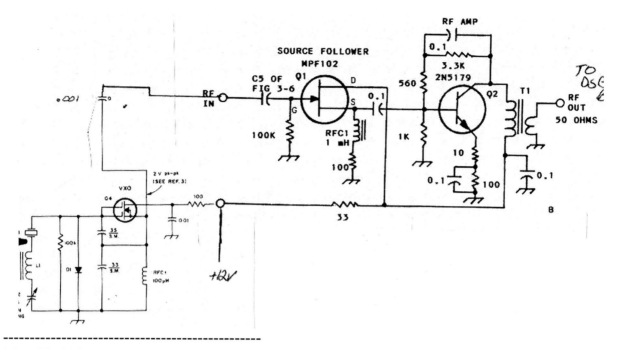

May 23, 2001

My 17 Meter DSB QRP rig has passed its smoke test. I built a two stage linear amplifier for it and it looks like I'm getting about 1.3 watts out. The transmitter is currently on 4 separate PC boards spread out on my operating table. Clip leads connect it to my 12 Volt Gel Cell battery. I'm using a Drake 2-B as the receiver.

17 meters has been in good shape. This morning I listened to a nice QSO between a VK7 (Tasmania) and a W6 (California). But so far I haven't made a contact with the new rig.

I was hoping that someone out there could point their beam at the Azores and listen for my low power DSB signal. My VXO tunes from about 18.110 to about 18.130. I can be on the air tomorrow (24 May) from 0530 UTC to 0730 UTC. Please let me know if you can listen for me during this period.

May 27, 2001

I made my first QSO with the 17 meter DSB rig, but it was more of a test than a QSO. After much fruitless calling I was forced to resort to a scheduled contact. CU2BD pointed his beam at me from the other side of Ponta Delgada (3km)... and still had trouble hearing me. He said the audio sounded OK, but very weak.

The problem is low output in the DSB mode. I think I've localized the trouble to the balanced modulator. This is a simple transformer/two diode circuit. It appears on page 139 of W1FB's "QRP Notebook."

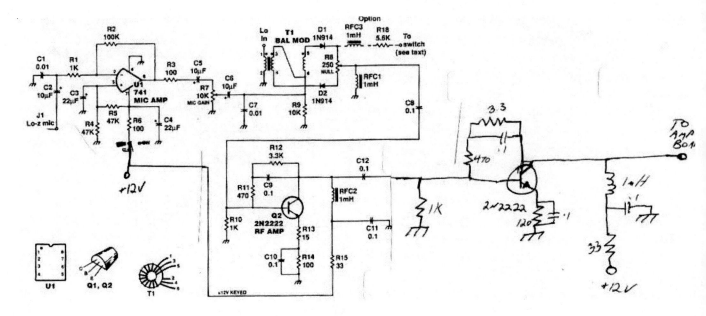

I am able to 'balance out' the RF. When I unbalance the modulator (using the balance potentiometer), I get about 1 watt out of the final (this is about right). But when I once again balance out the carrier and try to modulate, I can barely make the output meter wiggle. Even when whistling into the microphone, I can barely get the meter to move. I am modulating the signal, but just not enough.

I don't think the problem is insufficient audio. My audio amplifier (a 741 op amp) is working fine. And I tried injecting a 1 kHz tone (at about 1 volt rms) into the AF port of the balanced modulator—no joy. I put the same signal into the input of the audio amp. Again, no luck.

So it is as if half of the balanced modulator (the RF half) is working fine (I can null out the carrier) but the audio portion seems unable to "unbalance" the modulator properly. I think the audio should be able to produce a level of "unbalancing" similar to what happens when I turn the balance pot to either end, but so far it is not even close.

I'm using two Motorola IN5821977 hot carrier diodes. They both show forward resistance of about 5 ohms. (This is what I had in the junk box—do you guys think this is a suitable component?)

I'm stumped. Anyone have any ideas on what's wrong or how I should proceed?

May 28, 2001

With lots of help and advice from list members, I got the modulator working on my homebrew 17 meter DSB rig. I checked and re-checked everything and finally decided to pull out the Schottky diodes I had in there and try some ordinary diodes. (I don't even know the designation for the new ones—I picked them up at a Radio Shack a long time ago and they have been lurking in my junk box ever since.)

BINGO! With the new diodes I was able to null out the carrier and get approximately 1 watt out of the rig. I'm sure the audio doesn't sound great, but I can work on that later. For now, I'm very happy to see the power output/SWR meter bouncing nicely as I call (futile) CQs.

I don't know why the other diodes didn't work. They tested fine, with high reverse and low forward resistance.

Thanks again to all those who helped. I will continue to keep you posted on progress (and I will probably be asking for more help on future problems).

I think I'll be building another RF amp stage to get me up to 5-10 watts or so. Then it will be on to the companion Direct Conversion Receiver. This is, after all, a Direct Conversion transmitter.

June 3, 2001

After fixing my balanced modulator problem, this week I was trying to get the audio coming out of my 17 meter DSB rig to sound, well, human. There was clearly something wrong, something very nonlinear, in the RF amp that I'd cobbled together.

Yesterday morning, as I considered my options, I remembered that the 30 meter CW QRP rig that I'd built while in the Dominican Republic had a very nice, robust broadband driver and RF amp (two MRF476's in parallel). I haven't been using the 30 meter rig very much (I really hate using the direct conversion receiver that I built with it), so it didn't take long for me to convince myself to pull out the transmitter board and convert it into the power amplifier for my evolving 17 meter DSB rig.

It looked like all stages in the 30 meter rig (VXO-controlled 6 watter from QRP Classics) were Class A (linear amplifiers)—except, of course, the Class C final. I quickly built a little diode-based circuit to lift the bases of the MRF476's to a voltage that would keep things in Class AB. I disconnected the 30 meter output filter (I'll rebuild it for 17 later) and hooked the 17 meter dipole to the .1 capacitor that goes to the final.

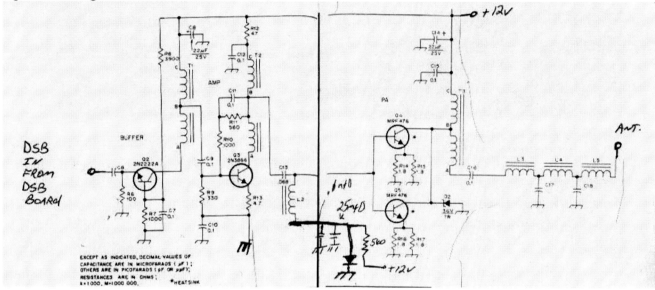

The driver transistor was running a bit hot so I improvised a heat sink out of a few inches of copper tubing and some heat sink compound.

I used the microphone from an old Yaesu memorizer transceiver. I had a bit of trouble getting the microphone to stay connected, so I whipped out some duct tape and stuck the connector and cables right to the desk.

Transmit-Receive switching is accomplished via the T-R relay that usually connects my Drake 2-B to my HT-37. So to transmit I must connect the DSB transmitter's ground lead to the Gel Cell and then put the HT-37 in transmit mode. This will all get easier, of course, when this rig is mounted in a cabinet.

This morning I had another schedule with Felipe, CU2BD, (in Ponta Delgada) to test the rig. I made the final connections with just minutes to spare. While I was waiting for Felipe to arrive on frequency, I heard CT2FYI calling from Lisbon. I gave him a shout, and got a good signal report (Lisbon is about 900 miles west of me!). Then Felipe came on and told me the audio sounded good. A few minutes later I worked D44BS in the Cape Verde Islands (I got a 55 report). Then came GB2RN on the HMS Belfast in London - he also gave me 55.

It was a lot of fun getting this thing to work. It looks like I'm getting about 2 watts PEP out. And I don't see any signs of distortion. I'm going to see who else I can talk to.

Next step will be to put the rig in a cabinet and build a more comfortable Transmit-Receive arrangement. Then I'll be looking for suggestions on Direct Conversion receive circuits.

Thanks again to all those who've been helping out with advice and encouragement.

June 4, 2001

In my last update I said that Lisbon was 900 miles west of me. I suppose that would put the Azores in the Med! I attribute the error to the early hour, solder fumes, and the excitement of getting the DSB rig on the air.

The transmitter continues to yield great contacts. Yesterday afternoon I called CQ on 17 and got a response from Jorge, EA5GQI/M, who was on the road in Alicante, Spain. Jorge was appropriately surprised when I told him I was running 1 watt from a homebrew transmitter. He pulled to the roadside, recorded my transmission and—from his car—sent my signal back to me. So I got to hear my own audio. Very cool!

Last night I again dared to call CQ and got a response from OZ4B in Copenhagen. Bo gave me a 55 report and said the audio sounded good. Not bad for an audio section that consists of a 741 op amp. Bo also suggested that I put the rig on 12 meters. Hmmm...

These contacts were great fun, but the big DX thrill came this morning. U.S. stations were coming in and I was hoping to work the homeland. No luck. But just as I was about to give up, I heard Gerry, VK7GK, on Tasmania say that he'd listen for one more. I gave him a shout. The DSB rig was 44 *down-under* down-under. We had a nice QSO.

When I was a teenage ham, the first time I'd worked a station in that part of the world, I woke up my parents to tell them of my feat. Good thing my wife was already awake, because I once again went bounding up the stairs with the big news. I'd been hoping to cross one ocean, and ended up crossing two.

On the tech side, I'm thinking that my decision to recycle the broadband RF amps from the 30 meter rig will yield a big, unexpected benefit: It will now be very easy to put this rig on other bands. All I'll have to do is get the appropriate crystal for the VXO and build a group of band-specific output filters for the final amp.

You guys have got to build some more phone rigs. This is really fun. You can hear the surprise in the voice of the guy at the other end when you tell them about the power level and the brand name (HB) of the rig. Phone QRP is a LOT easier than I expected, and I'm 10 db below the upper limit on output power for QRP status. People are answering my CQs!

When the dust settles, I'll draw up the schematic and post it (with pictures) on my web site.

June 18, 2001

While I have been having a lot of fun having QSO's with the three boards spread out on the operating table, the Transmit/Receive (T/R) switching arrangement was getting kind of tiresome. I've been using my Drake 2-B as the inhaler, and for T/R switching I've been using the external relay that is controlled by the T/R switch on my Hallicrafters HT-37. On receive I'd also have to shut down the transmitter by removing the ground connection to the Gel-Cell. Like I said, kind of tiresome. And frequency spotting required even more gymnastics! To make matters even worse, it was taking a few seconds for the electrolytics in the microphone amp circuit to get going, so there would be an annoying delay every time I went from R to T... you know, just long enough to have the other operator start calling CQ again. And the HT-37 was heating up the room. Clearly something had to be done.

I found a suitable 4 pole double-throw 12 V relay in the junk box and this morning I rigged up a much better TR scheme. The relay is superglued to the piece of pine that serves as the chassis. For the time being, one pole switches 12 volts DC to all the transmitter circuits except the microphone amp, which is powered up all the time. Another pole mutes the receiver and a third switches the antenna. I realize that the relay is a waste of energy, and that there are probably better ways of doing this, but this was a very simply way to allow for Push to Talk (PTT) operation. PTT was a major improvement.

Getting the boards screwed down to the chassis was another big step toward making this thing seem like a real radio. I have the DSB generator on the left and the small RF AMP board in the back center. The VXO box is elevated a couple of inches above the chassis by some pieces of wood—this will allow the main tuning cap to be at a comfortable position on the front panel. The board on the right side has been left open for the DC receiver. I'll post some pictures next week.

Following Doug DeMaw's advice about raiding the kitchen for radio cabinet material, I got hold of a couple of suitably-sized aluminum cooking pans. I should be able to make a good front panel from them.

So now my thoughts turn to the receiver (RX). Obviously direct conversion using the RF from the VXO.

I'm thinking about using a diode ring mixer for the RX (I hate AM breakthrough). I have to leave the VXO on all the time, and I'd like to avoid switching the VXO output from the RX to the TX (potential for freq shift under different loads) but I'm a bit concerned that adding another Low Z port to the VXO output might cause the input voltage at the TX LO input port to drop considerably. Any ideas on this? How about an FET isolation amp between the VXO and the diode ring detector? Or how about using one of those TV signal splitters at the output of the VXO?

I want to use discrete components in the RX. I think I want an RF amp stage (maybe grounded gate) just to get some additional tuned circuits between the mixer and the antenna. After the mixer I'll want some simple audio filtering (to get at least some selectivity). And enough audio to drive a small speaker. Any circuit suggestions?

June 28, 2001

I decided to use one pole of the 4 pole double throw TR relay to switch the VXO output from the transmitter balanced modulator to the receiver's mixer.

Looks like the receiver will be direct conversion with one stage of RF amplification, a diode ring mixer, a simple SSB low pass filter, and enough audio amplification to drive a speaker. (I'll probably resort to an LM386 integrated circuit chip.)

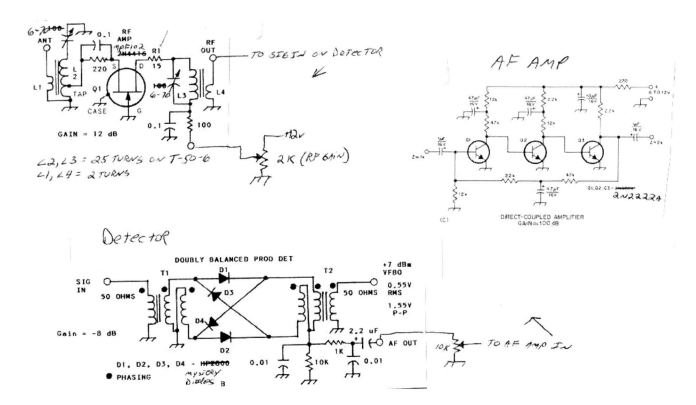

July 11, 2001

I've been having so much fun with the 17 meter DSB transmitter that I've been delayed in building the receiver. Let me give you an example of the 17 meter fun: On 5 July 2001 at around 0600 UTC I called CQ on 17 meters. With a regulated 13.8 volt supply the rig is putting out around 5 watts PEP (that's DSB, so there is a lot of energy in the other sideband and some in the not-fully-suppressed carrier). VK1MJ (Australia) came back. Mike gave me a 52, but I was very pleased. This was the second time in a month that he'd answered my CQ. When I finished with Mike, I called CQ again on the same frequency and was answered by KH8/N5OLS in American Samoa. I was 55. Don and I had a nice ragchew about island life (Atlantic and Pacific). My theory is that the radio gods smile on homebrewers.

So you can see why I've been neglecting the soldering iron. Fortunately propagation conditions deteriorated a bit during the last week or so—this allowed me to build a couple of stages on the DC receiver that will accompany the transmitter. First I built the four diode mixer. I used the same kind of small signal diodes that worked well in the TX's balanced modulator. Doug DeMaw's circuit called for the use of FT-37-43 cores for the transformers, but I'd found that core to be a bit small for my fingers, so I used FT-50-43 cores (same material, just larger diameter). I quickly connected the transmitter's VXO and tests indicated that the thing was in fact mixing.

Next I built a grounded gate FET transistor (MPF-102) RF amp with tuned circuits at the input and output. For the tuned circuits I used toroids and trimmer caps. Manhattan building techniques seemed to help a lot in keeping this amplifier from oscillating. Soon I had the amp connected to the mixer. From the junkbox I found an LM386 AF amp circuit that I'd built for an earlier project. That went to the mixer output. Now I had a receiver.

I think it is really cool when you first coax a signal out of a new receiver. Last night I was peaking the trimmer caps on the RF amp when all of a sudden I heard some very faint SSB chatter. I tuned the VXO, tweaked the caps a bit more and there it was, the unmistakable accent of a British radio amateur. I went to bed a happy homebrewer.

Now I have to build a real audio frequency (AF) amp for this RX. I'm tired of the standard LM386 or 741 IC op amp circuits. Isn't there something out there that is 1) simple 2) uses discrete components and 3) can drive a speaker? Any suggestions?

July 16, 2001

I added a 100 db audio amplifier after the mixer in the Direct Conversion (DC) receiver. It's the three transistor direct coupled amp that appears on page 76 of Solid State Design for the Radio Amateur (SSDRA). I used 2N2222 transistors. A board mounted 10K potentiometer went between the mixer and the input of the amp. I built it Manhattan style with some ugly thrown in. With high impedance headphones the amp works very nicely. You get that great sensation of hearing the band directly.

The high impedance headphones I'm using are very old. They were manufactured by "American Bell" in Wayland N.Y. I felt like a real radio pioneer when I put those things on.

I take back all the bad things I said about DC receivers. This one works very nicely. The VXO keeps it stable and the 4 diode mixer seems to have eliminated the AM detection problem (admittedly this problem is easier to beat at 18 MHz than at 7 MHz or 10 MHz).

Just to review, here's the receiver lineup: Grounded gate RF amp (MPF102) with tuned circuits in the input and output. Four diode mixer (ordinary switching diodes) with two toroidal transformers. The local oscillator is the VXO from the transmitter (an FET transistor oscillator, a source follower buffer circuit, and an amplifier). Then there is the audio frequency amplifier described above.

Question: In the mixer circuit that I used, at the mixer's audio output port, Doug DeMaw had a 10k ohm resistor to ground and a 1 K resistor between the transformer and the output cap. In another similar circuit, the output went across an RF choke to ground. Are these simple circuits in lieu of a diplexer? Do I need a diplexer?

I'm going to try to add one or two more audio stages to see if I can drive a speaker (without having this thing turn into an oscillator).

There is some annoying AC hum when I run this receiver off an AC power supply—if filtering and shielding doesn't fix this, I may have to limit myself to battery operations.

Last night I was tuning through 17 meters with the new RX and heard a loud CQ from ON4AAM. I hit the push-to-talk and gave him a shout. I was 56 - 57. I was very pleased with the first transceiver contact with this rig.

July 26, 2001

I'm having a lot of fun with the DC receiver that I built as part of my evolving 17 meter DSB transceiver. (I've posted pictures and a schematic at http://www.gadgeteer.us Just click on the 17 meter DSB link.)

Right after I built the receiver, I was browsing through some old *QST* magazines and came across the November 1968 issue. Wes Hayward, W7ZOI, and Dick Bingham, W7WKR, had an article entitled "Direct Conversion - A Neglected Technique." This must have been the piece that launched the DC revolution. Neglected no more! I knew I'd been on the right track when I discovered that the receiver described in the article was very similar to the one I'd just built. Same product detector and AF AMP. (This came as no surprise as I'd taken the circuitry out of Wes Hayward's book, "Solid State Design for the Radio Amateur.")

My previous experiences with direct conversion receivers were pretty bad—lot's of AM breakthrough, common mode hum, etc. I think I avoided many of these problems with this receiver by staying away from the standard NE602-LM386 neophyte designs. And I've been helped by the fact that earlier rigs were built for 10 and 7 MHz, where shortwave broadcast interference is much more intense than at 18 MHz. The AC hum problem was cured with one toroidal inductor between the power supply and the rig.

I have had some problems with AM breakthrough. On Sunday morning at 0800 UTC, Radio Exterior de España fired up, apparently with its antennas aimed in my direction. I consulted their schedule and I think I was hearing either their 17.77 or 17.88 MHz broadcast. Too close for comfort. Pretty much wiped out the ham signals I was listening to. I went back and very carefully re-tuned the two tuned circuits in the RF amp (front end). I found that I could peak the both of them in such a way as to eliminate most of the AM breakthrough problem. I know many people advise against radio frequency (RF) amps ahead of the mixers in DC receivers, but I wanted to get the two tuned circuits in there. I'm glad I did.

The RF amp can, however, give me too much gain and make things very noisy. I've seen people use a 10K potentiometer as an attenuator at the antenna port, but I came up with what I think is a more elegant solution: I put a 2k pot on the 12V line going to the MPF-102 in the front end. So I can control the voltage going to the drain of the grounded gate amp. This serves as a very smooth RF gain control. I use it a lot.

The receiver is stable and sensitive. Of course it is very broad and I hear "both sides." But I'm very satisfied. There is, of course, still room for improvement.

I'm still using the headphones. The receiver is working so nicely that I'm reluctant to add the extra stages needed for speaker operations. But I suppose I'll eventually give it a try. I don't even have an audio gain control. (The receiver in the November 1968 *QST* article didn't have one either. The authors suggested detuning the caps in the front end if you needed to reduce gain.)

One deficiency I can hear involves very strong SSB signals. Even if they are completely out of my tuning range, I can hear the "monkey chatter." Tuning the RX does not affect the sound. Same chatter no matter where I tune. Anyone have any thoughts on how to cure this? I don't know if the stations involved are splattering or not. I'm thinking that a low pass filter between the detector and the audio amp might help. I'd be looking at a simple, passive filter—I don't want to put in any complex active filter circuits. Thoughts?

I used some ordinary switching diodes in the mixer. Does anyone think that the receiver would work significantly better if I replaced them with the recommended Schottky diodes?

I'd been thinking that I'd need a lot of shielding around the RX board to make this thing work properly, but so far it is working fine with no shielding. The VXO is in a box, so I suppose that helps with some of the problems often seen in DC receivers. Anyone think I should put in some more shielding?

August 5, 2001

I've been trying to add more AF gain to my direct conversion RX for 17. I have a 100db, three transistor (2N2222a) direct coupled amp similar to the one from SSDRA. This amp works very well— it is very quiet and does not take off (oscillate) on me. I have a grounded gate RF amp (about 10 db) ahead of the mixer. I think there is about 8 db of loss in the mixer.

I know the RX would work better with more AF gain. When I plug a little amplified computer speaker in the AF out from the 100 db amp, I can turn the RF gain very low and the whole system seems to work better.

I tried adding an MPF102 amp stage ahead of the 100 db amp. It added far too much noise. Then I tried an additional 2N2222A stage AFTER the 100 db amp. Again, way too noisy. Today I built a 40 db amp out of the ARRL data book. Again, too much noise.

What am I doing wrong here? Am I proving that there is wisdom in the decision to go with LM386s and other ICs as AF amps? Is it really this difficult to get sufficient AF gain in a DC receiver using discrete transistors and simple circuits? Why is it that the 100 db amp is so quiet and works so well, while the others add so much noise?

BTW: I continued to have some AM detection problems with a Radio Exterior de España shortwave broadcast on 17.7 MHz. I put a series tuned circuit between the RF input and ground. A bit of tweaking knocked the SW signal down nicely while leaving 17 meters intact.

Don't get me wrong—the receiver is working nicely already. When I talk to Amadeus, CT2HGL, in Portugal, I can hear his parakeets in the background... Recent DX: 4Z4 (Israel), 9K2 (Kuwait), Shetland Islands....

August 27, 2001

Poor band conditions on 17 allowed me to put down the microphone and work on the rig.

Recognizing the inherent shortcomings in the receiver that consists of only 4 transistors and 4 diodes (other than the VXO), I decided to build a switching arrangement that would permit me to go to an external receiver (my trusty Drake 2-B) with the flip of a front panel switch. I used a three pole, double-throw switch and a small relay, and a second switch (to turn on the VXO for spotting). The circuitry works very well—when the DC receiver can no longer handle the interference, or when the weaker signals get lost in all the noise that is getting through the relatively wide passband of the DC receiver, I just throw the new "INT/EXT RX" switch, turn on the VXO, quickly zero beat the 2-B and I'm in business. (Of course, in keeping with Murphy's Law, the Drake 2-B died just as I completed the INT/EXT circuit! Thank God I had a spare rectifier tube in the junk box.)

Here's a question for the groups: Am I right in thinking that when the going gets rough (static, interference) the simple DC receiver will simply not be able to compete with a good superhet like the 2-B? I think the passband (and the noise) will always be about double that of the superhet, correct?

As for my audio amp problems, after many failed attempts to add additional AF amplification, I came across an article in QRP Quarterly magazine that discussed the big differences in sensitivity among high impedance headphones. This prompted me to reach into the junk box and experiment. I found a set of old military headphones that were significantly more sensitive than those that I'd been using. So I no longer have a need for more amplification. (I'd given up on the idea of having a speaker in the cabinet. If I want to listen on a speaker I can just switch to the 2-B.) With the high sensitivity phones, I needed to turn down the AF gain—so I put a 10k pot between the output of the 100 db pot and the headphones. (I didn't want to invite feedback by putting the long leads at the input to the audio amplifier.)

My problem with the overload of the audio amplifiers by mixer products beyond the audible range seems to have been reduced significantly by turning back the RF gain to minimal levels. Jake, N4UY, is sending me some 88mH toroids. I'll try these in the diplexer circuit to see if some additional low pass AF filtering will help.

Here's a transmitter question: Why did Doug DeMaw use a balanced modulator with only 2 diodes? Wouldn't he have gotten better carrier suppression with a 4 diode arrangement? (My carrier suppression doesn't seem to be that great—I can see about 250 mw of carrier when I push the PTT down. Output PEP is probably 5-8 watts.)

I put the rig in an old Bud cabinet and used the bottom of a cake pan as the front panel. I think it looks pretty good. Pictures on my web site.

UNDERSTANDING: BALANCED MODULATORS

Looking back at my e-mails on this project, it is obvious that I did not understand how that balanced modulator stage worked. In those messages I can see myself scratching my head about why DeMaw chose to use two diodes and not one or four, why certain diodes worked and others did not, why a certain minimum voltage level was needed. The balanced modulator circuit is really at the heart of almost all phone transmitters. Obviously, if I wanted to be among the anointed, I needed to know more than just what goes in and what comes out. As young Maxwell would have said, I needed to know "the go of it... the particular go of it."

This is a particularly pleasing circuit to analyze and understand. There are only a few parts in it, but many of the most important principles of the radio art are involved: In this circuit you can see Einstein at work in the transformer (and in the coil at the output). In the diodes you can

see the groundbreaking semiconductor work of Shockley, Bardeen and Brattain. With the one capacitor in the circuit, you can apply the words of wisdom from Asimov and Feynman. Fourier helps us make use of the jumbled signal we find at the output. And I think there is beauty in the symmetry of this circuit.

I think the best way to understand this circuit is to break it up into two parts, two actions: balance and modulation.

We've already covered the modulation part: it is mixing—it is what happens when you put two different frequencies through a non-linear device. That curved operating characteristic, or the completely non-linear characteristic of a switch, causes the output to be a complex periodic wave, and in that complex wave there are new frequencies: sum and difference frequencies.

We usually want to make use of these mixing products while leaving the input signals behind. The 'balanced" part of these circuits lets us do that.

Doug DeMaw's DSB rig had in it one of the most common balanced modulator circuits: two diodes fed by a transformer with a variable resistor at the output, and inputs for both RF and AF. RF and AF go into their respective input ports, and at the output we find the sum and difference frequencies (plus one of the input frequencies --- more on this later).

Here is how the circuit is often drawn:

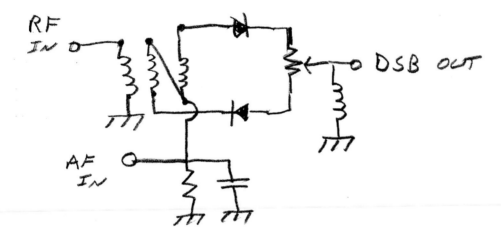

The three coils are really a trifilar toroidal transformer. (Say that three times fast!) You just take three pieces of insulated wire, twist them together, and then wind the three of them together onto a toroidal core.

I think this circuit is a good example of how sometimes it is worthwhile to re-draw a circuit. I think the circuit's functions are easier to see if you draw it this way:

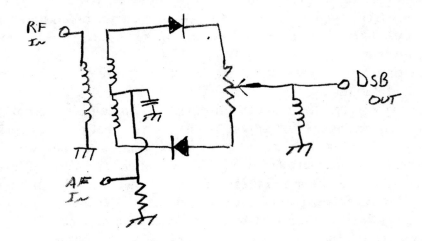

To see how the balancing takes place, let's take a look at this circuit without the diodes and without the variable resistor.

The RF signal comes from a local oscillator circuit. The current through the primary coil sets

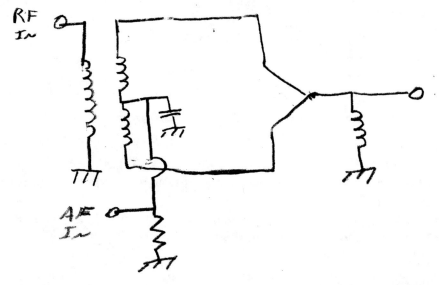

up varying magnetic field that induces an emf in the two secondary coils. You can look at these two coils as being one center-tapped secondary coil. When one end of this secondary coil is positive, the other end will be negative and vice versa. But at the output, these two oppositely charged points are connected. Positive and negative balance out, so there will be no RF at the output.

Now, look what happens to the AF. The AF signal goes simultaneously through both halves of the secondary coil, and is NOT balanced out at the output. You can see how the transformer, and how it is fed, determines which of the inputs will be balanced out. This is known as a "singly balanced" design—only one of the inputs (in this case the RF) is balanced out.

There are a couple of additional bits of magic in this part of the circuit. Because the audio is simultaneously going through both halves of the secondary coil (in opposite directions) it will set up equal but opposite magnetic fields. These fields will cancel in the primary, thereby improving the isolation between the two input ports (you won't have much AF energy going back into your RF source).

We need that center connection in the secondary to look like two different things to the two different input frequencies: For the RF, we need a ground connection (so that the two opposite ends of the secondary will be of opposite polarity relative to ground), but for AF this center tap needs to be the input port, and definitely should not be a ground. The capacitor connected to the center of that secondary makes this happen. It has a value that makes it look like a low impedance short circuit to ground for the RF, but for the AF it will appear to be a high impedance.

Now let's put the mixing diodes and the variable resistor back in, and look at how the mixing occurs.

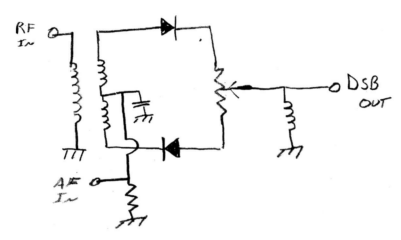

This circuit is really a "switching mixer" that operates much like the CMOS circuit of Leon, VK2DOB. When the RF signal goes positive, this will result in a negative voltage on the cathode of the upper diode and a positive voltage on the anode of the lower diode. Both diodes will switch "on" and will conduct, allowing the unbalanced AF signal to pass through. But when that RF signal swings negative, both diodes will turn off, and the AF signal will not get through.

In essence, the diodes will be "chopping up" that AF signal at the rate of the frequency of the RF input. It is important that the RF signal be of sufficiently high peak voltage to turn those diodes on. And it is also important that the audio input NOT be strong enough to trigger the switching.

The complex waveform that results from this electronic slicing and dicing will—as Fourier's math and Leon's diagrams predict—contain the sum and difference frequencies.

It is important to understand what will be balanced out by this circuit, and what will NOT be balanced out: Both diodes will simultaneously be "chopping" the same audio signal, so sum and difference frequencies generated in the top diode will be identical to those generated by the bottom diode. There will be no balancing out of these sum and difference output. But the RF input signal itself will be balanced out by the process described above. It is important to remember from our discussion of AM modulation that the carrier signal really remains unchanged in the modulation process—the sidebands pop up around it, but the carrier remains unchanged. And because of the "differential" way that carrier (the RF input) is fed to the diodes, it will continue to be balanced out at the output.

What about the AF input? Well, it will continue to NOT be balanced out. It will make it to the output. But because it is so far from the RF operating frequencies that we are interested in, it will be very easily filtered out. In fact, that coil at the output can be seen as the first step of the filtering process that eliminates the AF—that coil will send much of the AF frequency energy to ground, eliminating it. But that same coil will NOT send the sum and difference energy to ground—those frequencies are at RF and will continue on to the amplifier circuits.

There are balanced modulator designs that balance out BOTH the inputs and leave only the mixing products. Those circuits often employ four diodes in a ring configuration. But there is no real need to balance out the audio—it is easily taken out of the picture by filtering action. That's why DeMaw's designs used two (not four) diodes. You need at least two (to balance out the RF) but with these frequencies, four would be overkill.

In the course of trying to understand DeMaw's two diode circuit, I came across a book that really helped me understand how balanced modulators work. I found it in an unlikely place: a RadioShack store. In spite of its seemingly very "ham radio" name, over the years RadioShack has become the target of unjustified scorn from technically oriented hams. Maybe it's because they sell CB gear. Maybe its because the salespeople sometimes might not be real technical experts... For whatever reason, RadioShack seems to get a lot of criticism in ham circles. But I've always liked their stores, and I've always missed them when overseas. And one of their books really helped me with the balanced modulators.

RadioShack's book, *Basic Communications Electronics* by Jack Hudson and Jerry Luecke, has (on page 94) this wonderful diagram:

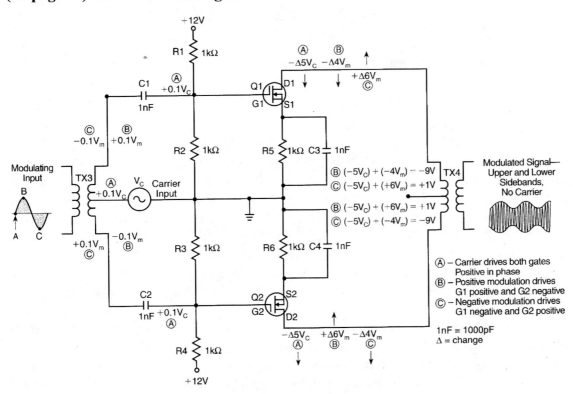

This one is a bit different from DeMaw's circuit, but the underlying principles are the same. Doug's circuit was a switching mixer—like the CMOS gates in Leon's circuit, the diodes did their mixing magic by operating in that far extreme non-linearity: on or off mode. This RadioShack circuit is not a switching mixer—it relies on the curves in the operating characteristic of the two FET devices.

In their circuit description, the authors look at what happens in the circuit at three different points in one cycle of the modulating signal. For simplicity sake the instantaneous voltage of the carrier frequency is assumed to stay at +.1 volt in each of the three cases.

At point A there is no modulating input. Because of the way the input transformer is set up, identical +.1 volt carrier signals are applied to each gate. Both of the closely matched FET's respond in the same manner, and both of them put a -5 volt signal onto their ends of the primary of the output transformer. But hey, with the same voltage at the same polarity on both ends of the transformer, there is no potential difference there, so there will be no current through the primary and no output at the secondary. The carrier signal is balanced out.

Now lets look at what the situation would look like if the modulation signal were to be at B—a positive peak of +.1 volt—when the RF was at +.1Volt. This modulation signal means that the gate of the top FET goes from +.1 volt to +.2 Volts. You'd think that this would result in an additional – 5 volts at the top of the primary of the output transformer, but—and this is very important—we have to remember that these devices are deliberately set up for NON-LINEAR operation. There is NOT a linear relationship between input and output. That operating characteristic is curved. Doubling input voltage DOES NOT result in a doubling of output voltage. In this case we see that the additional +.1 volts at the gate only results in an additional – 4 volts at the top of the primary of the output transformer, momentarily putting a total of – 9 volts at that point.

And look what's happening in that bottom FET. Because of the way that input transformer is configured, unlike the situation with the RF input, the modulating audio input puts voltages of OPPOSITE polarities on the two gates. So at the same moment that the top gate is getting an additional +.1 Volts, the bottom gate is getting -.1volts from the modulating signal. This -.1 volt modulating signal would cause a +6 volt signal to appear at the drain. (You might have expected this voltage to have put a +4 volt signal on the output, but remember, these devices are non-linear.) Combined with the – 5 volts resulting from the carrier signal, we now have +1 volt on the bottom end of the transformer. – 9 volts at the top, and +1 volts at the bottom represents a potential difference of 10 volts. Current flows.

The non-linear operating characteristics of the FET circuits creates a complex output waveform that—as Fourier and the trig equations tell us—will contain the sum and difference frequencies. The way the transformers and the inputs are arranged makes sure the RF carrier signal is balanced out.

To make sure you understand what is happening in this circuit, ask yourself this question: What would the output signal look like if the FETs were linear?

Here's the answer in Morse code: .- ..-.

(I'm still not sure why the original hot carrier diodes didn't work.)

****

That little 17 meter homebrew rig soon turned into an electronic equivalent of a magic carpet. At this point we were past the peak of the sunspot cycle, but there were still enough freckles on Old Sol to keep the ionosphere reflective. Although my new creation put out only about as much power as a small night-light, my very favorable position in the North Atlantic meant that even with modest antennas I was regularly able to contact stations in Continental Europe, North America, and in distant Oceania.

One of the first friends I made with the new transceiver was Maurice Newell, G3IUE. I was calling CQ one day, and Maurice came on to tell me (very nicely) that I was transmitting on the *wrong* sideband. I explained that I was transmitting on both. As a long-time homebrewer Maurice got a real kick out of that. He had a wonderful location for radio work—I'd frequently hear him talking to other stations, telling them that he was transmitting from "Penzance, near Land's End." Maurice was in the far southwest corner of England, in a beautiful area that actually supports a few palm trees.

Maurice and I spoke frequently. I would often hear his wife speaking in the background. Occasionally I would hear the seagulls outside his window. Maurice was 74 when we met on the air. I knew from his occasional reference to pain killers that he was suffering from something long-term, but he never talked about it too much, and it was obvious that he never let it get him down.

From his Land's End shack, Maurice was involved in a high-seas adventure. A radio amateur from the Falkland Islands named Shorty, VP8NE, had decided to embark on a sea voyage of global scale—he was sailing from his home in the Falklands, up to Norway, and back. Demonstrating a truly remarkable dedication to the art of technical modification, Shorty had taken a small ship, cut it in half, and extended its hull to 20 foot length. Then, with almost no sailing experience, he and his girlfriend Allison had set out to sea. Baby Thomas had been born en route—"then we were three" noted Shorty. In one of our contacts I could hear the kid in the background as his dad spoke to me on 17 meters from somewhere out in the Atlantic. Thomas would take his first steps aboard the small ship.

Maurice was a friend of Shorty's and kept a regular schedule of radio contacts with him, monitoring progress as he, Allison and Thomas made their way up through the Atlantic. We followed their journey up the coast of Brazil then across to Europe, through Loch Ness, and on to Norway. They are now safely back in the Falklands, and young Thomas has learned how to walk on land.

The 17 meter band seemed to be filled with people like Maurice and Shorty. Radio amateurs had acquired 17 meters during the 1980s. There was an international agreement to keep these new frequencies free from the kind of amateur radio contests that often turn the other bands into hyper-competitive chaotic tangles of mutual interference. "Seventeen" was very un-competitive, very relaxed and friendly. And it had wonderful potential for long-distance communication. It often seemed that seventeen meters served as a sort of electromagnetic watering hole for a group of congenial and interesting radio amateurs.

This watering hole was important to me. Out on the island, it was always difficult for us to put aside our official role. Just about everyone always ended up treating me as O Consul (The Consul). But on 17 meters, nobody cared about that. 17 meters represented for me an important opportunity to reach out (way out!) and make friends. It was a lot like the 75 meter AM frequencies that I'd listened to as a teenager.

Rolf Schick, DL3AO, was definitely in the "congenial and interesting" category. He is a retired vulcanologist living in Stuttgart. He had spent his professional life studying the world's volcanoes, and his travels had at one point carried him to Sao Miguel island. So, in addition to our common interest in radio, we often spoke about volcanoes (including the one that I was sitting on). Rolf is the author of *The Little Book of Earthquakes and Volcanoes* and *Volcanoes of Europe*. During one contact Rolf told of an upcoming visit to volcano-plagued Montserrat Island. He talked of climbing Mt. Etna with his son. And of course, we talked of homebrewing radio gear—Rolf had built his own single sideband rig back in 1959.

Percy Masters, EI9FN, lives in Galway, Ireland. Percy had a massive wire "Vee-Beam" antenna that happened to be pointed right in my direction. Percy's signals were always booming in. In addition to talking about antennas and DX, Percy and I talked about languages. He is very fluent in Spanish, and knows all about Spain's other languages: Catalan and Euskera.

Ed Stokes, WB6KOK, lives in Maine. For his antenna, he operated a large array of vertical radiating elements, with the energy to each one phased so as to radiate a strong signal in a selected direction. When he aimed at Europe or the Middle East, I was in the path of his signals. We had many great conversations. It was good to talk to the homeland.

Chris Brannas, SM0OWX, is a 17 meter devotee who lives in Stockholm. When conditions permitted, I was talking to Chris several times each week. His friendly voice calling CQ seemed to become part of the frequency band.

Lindsay Britton, VK3CML, embodies the rugged individualism of Australia. He lives in a place called Stawell (pronounced "Stall") a town of 7,000 people in North-West Victoria State. Lindsay operates a homebuilt linear amplifier and uses a massive six element log-periodic antenna. He sent me photos of his backyard. Kangaroos were out there among the antenna towers.

I first met Mike Hopkins, EI0CL/EA8, on December 30, 2001. Mike is an Irishman, but he was at his Tenerife Island summer home when we first met. Mike is an amazing guy. Conversations with him would sometimes include discussions of astronomy, airplane construction, Wild Boar hunting, transatlantic sailing, and armed encounters in the Omani dessert (during one of which he first met his wife). For a while, Elisa refused to believe that Mike was for real. I'd tell her about what we'd been talking about and what Mike was up to, and she'd put on a suspicious face and express doubt that anyone could be involved in all the things that Mike is. A true radio amateur and experimenter, Mike demonstrated admirable interest in the ionosphere by naming his daughter "Aurora." When we first met on the air, he told me of having sailed his own boat out to the Azores. Sometimes, Mike and I would get on 17 meters and have an hour-long rambling conversation. After we finished, other stations would often call in to tell us how much they'd enjoyed listening in. This listener feedback gave me an idea that, years later in another country, would lead to the SolderSmoke podcast.

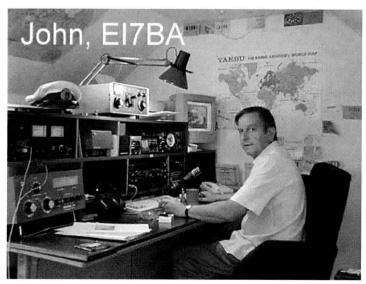

Mike is a friend of John Tait, EI7BA. John had been to sea, and knew Sao Miguel island very well. We could, for example, discuss some of the stranger items (two headed calves!) on display at the town's somewhat eccentric museum. A typical log entry for a contact with John showed us talking for over an hour, and covering steam engines, pirates, the British tradition of amateur science, the Wright Brothers, and the new 60 meter amateur band. John has definitely kissed the Blarney stone; he would often crack me up with very picturesque descriptions of technical things. For example, when we were discussing the miraculous properties of the WD-40 (oil in a spray can), John agreed that it was great stuff, but the way he put it was rather special. "It's the Pope's pee!" he declared. I'd never heard it praised in quite that way. When I was testing my new transmitters, he would look at them on his oscilloscope and report "no uglies." He had an antenna tuner that he claimed could tune up "everything from cuff links to the national power grid."

Rolf Lasses, SM0FQW, lives in Stockholm, but goes up north to a vacation home near Mora, Sweden during the summer. Rolf has a pet pig named Budde. We usually spoke on 17 meters, but sometimes we'd augment our connection with a Voice Over Internet link on the Echolink system. Rolf would connect by a small VHF radio to a local two meter repeater system that would pass his voice onto the Echolink system on the internet; in the Azores, his voice would be coming out of both my computer and my little homebrew transceiver.

As interesting and fraternal as 17 meters was, in 2002 there were some new things in the sky that caught my eye and started to divert some of my attention away from that friendly band. MIR was gone and the P3D satellite was severely disabled, but by now the International Space Station (ISS) was in orbit, and it had amateur radio equipment aboard. Much of the amateur gear was digital. In a similar, but smaller-scale development, Midshipmen at the United States Naval Academy at Annapolis had built their own satellite, dubbed PCsat; on September 30, 2001 they had sent it into orbit from the Kodiak, Alaska launch site. Of course, I found this all very interesting. With "PCSat" and the Space Station we had two digital repeaters in the sky. They could be used to relay packets of digital information from amateur radio ground stations.

The internet is a large scale digital packet system. E-mail, photographs, videos, spreadsheets, and our voices, are all digitized, broken up into packets, and sent out onto the net en route to their ultimate destinations. Years before the internet took off, radio amateurs were using digital packet technology to communicate. Instead of using the telephone lines and fiber optic cables that today carry the internet's traffic, radio amateurs sent their digital packets over the airwaves. To do this, a kind of modem known as a terminal node controller (TNC) is placed between a computer and a radio transceiver. The operator types his message on the computer, the computer and the TNC convert the text into digitized packets, and the transceiver sends the packets (which by then sound like digital electromagnetic burps) out onto the airwaves.

I had dabbled in packet radio, but never found it very appealing. My preference for voice operation was clearly a part of this, and I suppose I was also bothered by the very "appliance operator" feel of packet radio. The technologies involved are devilishly complex—there seemed to be little room for homebrewing in this microprocessor-intense area of the hobby. And when the internet burst onto the scene, and there seemed little point in using the amateur digital radio network when messages could be passed more easily and with greater reliability over the 'net.

But when digital circuits started to go into orbit on amateur satellites, I found myself dusting off my old TNC, and giving thanks that I had not sent it off to eBay. With PCsat and the ISS in the air, I began to construct what ended up being very much a junk-box satellite ground station.

I had last used the TNC with the old '286 computer that I'd acquired when I was living in Honduras in 1988-1989. This was the computer of the Contras. This computer was old. Old and rickety. The hard drive was starting to get a bit hit-and-miss. This was understandable—these things weren't supposed to last forever. But it was on this hard disc that I had the software to run the TNC, so I had an interest in putting it back into service.

For my radio uplink to the satellite I would use the old two-meter hand-held transceiver that in the Dominican Republic had allowed me to speak through Internet Phone while walking in the park with Elisa. At a Northern Virginia hamfest I'd picked up a small RF amplifier for two meters. This too was put into the project.

For an antenna, I resorted to the use of a coat hanger. This is an old trick for two meter antennas. You get a wire coat hanger or two and chop them up into five wire elements each of about 19 inches in length. Then, using a coaxial cable connector as your base you fashion them into what is known as a quarter wave ground-plane antenna. It is a simple, omni-directional antenna. It lacked the directionality (and the impressive appearance!) of my old Ray-gun antenna, but I wouldn't have to worry about where it was pointed—it was radiating RF in all directions.

The radio part of the station was coming together nicely, but the old computer was still very shaky. Some messages to the appropriate internet groups yielded a number of suggestions. One noted that the electric motors on the hard drives of these old computers had a tendency to seize up. This sounded fatal, and the old computer was halfway to the dumpster when I thought of WD-40. Ah yes! The fountain of youth for aging machines! The elixir of life for old electronics! The Pope's Pee! With nothing to lose I opened up the old case, found the hard drive's motor and gave it a judicious squirt.

It worked. The computer booted right up and began to display all the files and programs that had been left hidden by the balky drive. There was PACTERM, the program that linked the PC with the TNC. And there was Orbits II, the satellite-tracking program that had revolutionized our satellite program in the Caribbean. Putting all that old gear to work and using WD-40 on that hard drive made this project seem a bit more homebrew, a bit less plug-and-play.

On Thanksgiving 2002, I downloaded the most recent Keplerian elements for the space station and loaded them into the old Orbits II program. I noticed that the space station would that night be passing overhead shortly after dusk. This was the perfect scenario for visual observation: Two hundred miles up, the very large station made of shiny aluminum would still be in the sunshine, while we on the surface would already be in the earth's shadow, in darkness. And I knew that this would be a particularly good pass to watch, because there was some additional reflective material up there that night: the Space Shuttle Endeavor was docked with the space station.

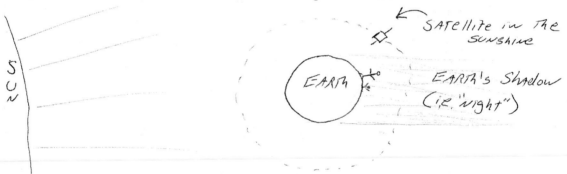

As the hour for the satellite pass approached, I called my friend Messias and gave him the frequency and the time of the pass. Messias had been using packet radio for many years and was well equipped to listen for the space station. I then turned on my computer, TNC, and transceiver, picked up five year-old Billy, and headed out to the backyard.

We knew it would be coming in from the Southwest. There it was, right on cue. Billy and I watched in fascination as the huge station soared overhead, moving at 17,000 miles per hour, then suddenly disappeared as it passed into Earth's shadow.

Back in the shack, we saw that my old Rube-Goldberg ("Heath-Robinson" for the Brits) satellite ground station had succeeded in capturing amateur messages that were being relayed by the digital repeater on the space station. The technical achievements added to the sense of contentment produced by all the turkey and mashed potatoes.

Of course, the next night we were at it again. At the hour indicated by Orbits II, Messias and I were both tuned in to the 145.800 MHz downlink frequency. Right on time, we started hearing the "burps" of packet signals coming down from space. A few seconds later, when the station rose a bit higher in the sky and got a bit closer, the signal got strong enough for our TNCs and computers to decode them and display the messages on our computer screens. We were astonished by what we saw.

The night before, we'd seen the callsigns of many continental European stations, and those of a few North Americans. But on this second night of downloading from space, we saw in the list of stations transmitting through the orbital digipeater the distinctive callsign of an Azorean radio amateur.

The Azores region has a population of only about 240,000 spread out over nine mid-Atlantic islands. When we started our satellite adventure, Messias and I both thought for sure that we were the only people in the region to be engaged in the very esoteric activity of listening to digital amateur radio signals from the International Space Station. But now we knew that we were not alone in this madness. There was another Azorean radio fiend out there, capturing packets from outer space. And he was launching a few of his own.

The call sign was CU3GC. CU indicates an Azorean station, the numeral 3 indicates Terceira Island. The numerical designation and the name Terceira ("Third") derive from the fact that Terceira was the third of the Azorean islands to be discovered. Messisas and I were both CU2s because Sao Miguel Island was the second to be found.

By using some online databases, we were able to find out who CU3GC was. It was Ray Mottley. He was an American assigned to the US Air Force facilities on Terceira. His U.S. callsign is WL7CDK (that one indicates Alaska—Ray obviously gets around). The database gave us an e-mail for Ray, and I quickly sent him a note reporting that his signals had been heard on Sao Miguel. Ray wrote back in astonishment reporting that the signals we had picked up were his very first efforts to pass a signal through the orbiting digipeater. Think of the odds of that. Two Americans living in the Azores, each on a different island, both decide to build systems that would permit them to communicate

through the space station. On the very first attempt, the American on Sao Miguel hears the American on Terceira. It was definitely eerie.

Of course, being true radio amateurs, we now wanted to make a two-way contact. After some e-mail coordination, we were ready. On November 29, 2002, as the International Space Station rose over the North Atlantic, I started sending packets through its digipeater. "CU3GC de CU2JL." Suddenly came a response from Ray. He told me that he was seeing my packets—I confirmed that I was seeing his. That made it a contact. Ray went on to note that we now owed each other QSL (confirmation) cards, and suggested that our Azorean satellite adventure be written up for *QST* magazine. (I followed up on that suggestion. See the September 2003 issue.) After we finished, realizing that Messias was monitoring, I gave him a call: "CU2BJ de CU2JL." I knew that he wasn't set up to transmit, but I knew that he would be thrilled to see his call coming down from the International Space Station. He saw my message and got a real kick out of it.

Messias and I had become good friends. I very rarely went to the club meetings—I was living on the schedule of my small children and the radio club meetings didn't start until after our bedtime. So I didn't actually see Messias all that often. But about halfway through my time in the Azores, Messias and I started meeting each other most evenings on the two meter band.

I was using the same old Yaesu Memorizer that my dad had given me many years before. This was the rig that I'd modified to work Morse Code through the RS-10 satellite, the same rig that had put me in contact with Norm Thagard on the MIR station, the same rig that I'd hooked up to the Internet Phone VOIP system in Santo Domingo. By now it was really starting to fall apart. The final audio amplifier transistors in the receiver section were burned out, and I didn't have anything to replace them. I slapped together a very inelegant solution by putting in a small jack that allowed me to substitute an amplified computer speaker for the rig's defunct audio amplifier. It wasn't pretty, but it worked. All the history and hacking and modification made that Yaesu much more than a mere appliance. My cell phone works perfectly every time, and I've never opened it up to fix it or modify it—so I'll never have the kind of connection to the cell phone that I have with that old and very aptly named "Memorizer." And now that old rig became my link to Messias.

Messias would leave his gear tuned to our pre-arranged frequency. At around 8 pm, I'd give him a call. He'd always ask how the kids were doing. In the English-speaking world, radio amateurs sometimes refer to their kids as "harmonics": smaller signals generated by a larger signal. But in the Portuguese-speaking world, the ham radio slang for children was "cristalinhos" or "little crystals." Azoreans love children, and Messias would always ask about my cristalinhos. As I talked to him, I could hear his wife Natalia working in the kitchen—I knew she could hear me so I would always say hello. Sending digital packets through the space station was fun and technically interesting, but those high tech digital signals would never be able to convey the kind of warmth and friendship that were carried by those simple, old fashioned analog voice signals.

But there was no denying that the satellites were a lot of fun. For me, they also served as a kind of link between amateur radio, and another hobby, amateur astronomy. With aspirations of mid-Atlantic astronomy, perhaps drawing sub-conscious parallels with the observatories on Hawaii's Muana Keia, and Spain's Grand Canaria, I'd brought with me to the Azores the 4.5 inch Newtonian reflector that Antonio and I had used in Santo Domingo. I'd pictured myself taking advantage of a big backyard and the pollution-free skies of the North Atlantic, using my telescope to explore planets, star clusters and perhaps the occasional distant galaxy. I soon discovered, however, that the Azores is one of the world's worst places for astronomy, and one of the world's best places for observing the undersides of gray clouds. Each morning before dawn, as I headed down to the basement shack, I would peer out the window. "Yep, clouds again."

But sometimes, of course, the skies would clear. I always had the telescope at the ready, and if the break in the clouds coincided with an official Consulate dinner, I'd set up the telescope in the backyard and invite our guests out for an after-dinner look at Saturn or Jupiter or some other eye-popping, jaw-dropping celestial object. This was always a lot of fun, and helped, I think, keep whatever political or economic problem we were discussing in its proper perspective. It is hard to get too worked up about the politics of the Azores (or of the U.S.!) after contemplating the enormity of the Andromeda galaxy.

Andromeda beckoned, but I kept getting pulled back to the interesting cast of characters on 17 meters. My little low-power double sideband transceiver was doing fine, but I could tell that the other stations often had to strain to pull my signals out of the noise. I wanted to make things easier for them, so improvements in the antenna and transmitter area were called for.

In amateur radio, there are guys who love to work on antennas. I'm not really one of them. I prefer a comfortable workbench or operating position to a windswept roof. And I have decided—for the sake of my children—to never risk life and limb by climbing an amateur radio antenna tower. So, continuing with the technique that I first used in my backyard in Congers, N.Y. I had largely limited myself to wire antennas suspended by tree branches. But of course, I often found myself yearning for more…

During our first year in the Azores my antenna work left me intimately familiar with all of the features of the roof of our too-large house. I had squeezed out of all the possible suspension points the maximum possible height for my wire dipoles. But I wanted more.

I started to cast a covetous eye at the highest chimney on the roof. It was way up there, a good thirty feet off the ground and above the surrounding buildings. It would allow an antenna to see the horizon in almost all directions. I pictured my homebrew radio waves skimming out along the water heading out past the horizon for their first big bounce off the ionosphere. For long-distance contacts you want your signals going out at low angles.

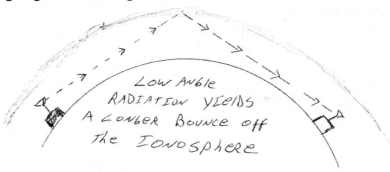

But that chimney was already in use. There was an old TV antenna up there. We'd long since switched to cable, and the satellite dish that we used for the Armed Forces television (Sesame Street and Oprah!) was on the other side of the roof, so I began to draw up plans to take the high ground.

Most amateurs would at this point be thinking of putting a large, directional array up on that roof. This would involve taking two or more dipole-sized antennas, making them out of aluminum tubing instead of wire, mounting all of them on a boom, connecting the whole array to a remote-control motor (to point it in the desired direction) and then putting the whole thing up on a tower.

Of course, that would have been great, but for a number of reasons I did not go that route. My homebrew instincts had been getting stronger, and I wanted an antenna that I could build myself with readily available materials. I also wanted something that I could put up relatively easily (less time up on the roof). And I needed something that would hopefully keep "Have-you-seen-that-thing-on-the-roof-This-Consul-is-REALLY nuts" comments to a minimum.

So I compromised. I'd go for an antenna that would combine some of the fun features of a rotate-able array with the simplicity and discretion of a humble dipole. I would build a one element Yagi. A single element beam. A rotate-able dipole.

As often happens with homebrewers, the components for this project had been gathering even before the idea was hatched. Several weeks earlier I'd been in the local fishing tackle shop and had spotted some 15-foot telescoping fiberglass fishing poles. Somehow, they looked useful. I bought five. "What are they for?" asked my wife. (A reasonable question when someone who does not fish suddenly purchases five large fishing poles.) "I'm not sure," was my response. But soon I would know…

I arranged for the obsolete TV aerial to be removed, and for a sturdy steel pipe to be fixed to the chimney. This would be the base on which my new antenna would be constructed.

Two of the fishing poles would be used to suspend the two wire elements of the antenna horizontally in the air (in the ether!). The wire would be held close to the fiberglass by simple cable ties.

In the Azores, one of the best perquisites that came with my job was the opportunity to ride around the island with Mr. Francisco Silva. Officially, Mr. Silva is the Consul's driver. But he is much more than that. For my family and me, he became a close friend, advisor, confidant, and at times, bodyguard. Mr. Silva had no interest in radio. His passions are the Benfica football (soccer) team and raising parakeets (a number of which were inadvertently released into the wild by my children). Mr. Silva knows his island like the back of his hand, and whenever I needed something for my radio projects, I would ask Mr. Silva where to get it.

For the antenna project, I needed a piece of wood to serve as the center support for the antenna. I mentioned this one day as Mr. Silva and I were leaving the town of Ribeira Grande, en route back to Ponta Delgada. He immediately put our armored Crown Vic into a turn and headed for the local sawmill. A quick word with the owner followed by a couple of minutes scrounging (in suits!) in the scrap heap produced the needed piece of wood. (This scrounging undoubtedly added to the already large collection of "Isn't the Consul a bit odd?" comments.)

Varnish was applied to the scrap lumber. The fishing poles were attached to the wood with U-bolts. Two other U-bolts would hold the wood to the mast that would rise from the chimney support.

Most of the radiation from a dipole antenna is, as you would expect, off the broadsides of the antenna. Very little goes off the ends. A dipole is essentially a bi-directional antenna. This makes it useful to be able to rotate the dipole—you can make sure that one of the radiation lobes is over the guy you are trying to talk to. And if there is interference, sometimes you can put the offending signal off the end of your dipole.

That's the electromagnetic reason for making the antenna rotate-able. But there is also a physical, mechanical reason to do this. IT IS JUST REALLY COOL to sit in your shack and make that big thing on the roof spin around! Between the fishing pole antenna and the chimney mast, I had installed an old TV rotor (the same one that I'd used with the Ray Gun in Santo Domingo). Now, next to my radio gear, I had this box with a big control knob on it marked "N-S-E-W." A twist of that knob would produce the sound of electrical motors in operation. The indicator would be moving to the desired heading, and up on the roof, my homebrew, single-element, fishing pole bi-directional beam antenna would be moving in step with the indicator, like a big electromagnetic eye-ball, turning its focus to another part of the globe. It was very cool.

The new antenna was an improvement, but my low-power double sideband transmitter still made contacts with me difficult for the guy at the other end. I was still transmitting less power than is consumed by your typical night-light. And because I was using double sideband, at least half of the power I was transmitting was being wasted; the receiving stations were all using single sideband receivers—they were only listening to one of the two sidebands. And because the new antenna was bi-directional, half of the power that I was radiating was not going towards my interlocutor. In sum, CU2JL was definitely not in the "easy-listening" category. Friendly, comfortable long-distance chatting was what 17 meters was all about. Clearly, something had to be done.

It might seem that an amplifier would be the solution, but there was a moral issue here. Double sideband uses twice the amount of electromagnetic spectrum that a single sideband signal consumes. There are no rules against using this mode, but it is kind of frowned upon by the amateur fraternity. The ethical problems could be minimized in a number of ways: I was cut a certain amount of slack because I was using homebrew gear—this was seen as highly laudable, and caused many people to overlook the spectrum issue. By using a direct conversion receiver, I would be listening to a stretch of the spectrum as broad as the one that I was transmitting on. This would help me avoid interfering with stations that were close-by in frequency. Most importantly, by keeping the power level of my double-wide signal low, I could minimize the havoc—and the complaints—arising from my "broadband" signal. Slapping a big amplifier onto that DSB transmitter would, I feared, make me something of an electromagnetic pariah.

So it was an issue of spectrum ethics that caused me to move from double to single sideband, to make a technological leap into the early 1960s.

I think all true radio amateurs are packrats. We come across odd bits of technology and literature and—without really knowing why—we tuck them away for future use. I'd been doing this for many years, and so had with me in the Azores a rather impressive "junk box" and ham radio library. This collection of radio rubble became the primary source of parts and ideas for my single sideband project.

In addition to the mainstream ham radio magazines like *QST* and *73*, there are now many specialized publications focused on niche areas of the hobby. One of the best is *SPRAT,* the quarterly journal of Britain's radio club for enthusiasts of low-power communication, G-QRP. G-QRP and *SPRAT* are part of Britain's long tradition of serious amateur science. Most of the club members are talented and experienced radio builders. Filled with innovative circuits, practical tips, and beautiful hand-drawn diagrams, the magazine oozes technical skill, electronic craftsmanship, and ham radio brotherhood.

In the fall of 1986, *SPRAT* had published a short article by Frank Lee, G3YCC, about a simple single sideband transmitter that he had designed and built. Frank's design provided me with inspiration and ideas.

A single sideband rig starts off very similar to a double sideband transmitter. First, a radio frequency oscillator and a microphone amplifier feed signals into to a balanced modulator. The two signals mix producing both sum and difference products aka "the sidebands." In the happy and simple land of DSB, the only thing that remains to be done is amplification and then it is off to the ionosphere. But with SSB, significant work remains.

In an SSB transmitter one of the two sidebands has to be eliminated. As we've discussed, there are several ways to do this, but the simplest is by using a filter. Just as the double sideband transmitter was the transmit equivalent of a direct conversion receiver (both circuits would allow both sidebands to pass), a filter type SSB transmitter is the equivalent of a selective superheterodyne receiver (both circuits filter out one of the sidebands).

A SINGLE SIDE BAND TRANSMITTER

As with the superheterodyne receiver, the filter in an SSB transmitter must be quite sharp. In the transmitter that I eventually built, the filter would be confronting two sets of signals, one around 5.174 MHz, and the other around 5.172 MHz. These were the sum and difference products that resulted from mixing the audio from my voice with the 5.173 MHz signal generated by the RF carrier oscillator. The filter's job was to allow one set of signals to pass, but not the other.

A double sideband rig starts out with an oscillator running at the desired operating frequency. Circuitry is included that allows this oscillator's frequency to be varied so that the transmitter can be placed at the desired spot within the amateur radio band in use. But the filter in an SSB transmitter makes this kind of simple tuning impossible. Filters usually need to be built for one specific frequency. So an SSB rig starts out as a fixed frequency transmitter. The carrier oscillator is crystal controlled, with no variation in frequency. The two sidebands that come out of the balanced modulator are therefore developed above and below this fixed carrier frequency. The carrier itself is eliminated by the balancing effect of the modulator. The filter slices off the undesired sideband, and sends the other off for further mixing and amplification. The filter is the core of an SSB transmitter.

Coming out of the filter you will have a nice, spectrum efficient single sideband signal. All of the necessary tones of your melodious voice will be in there. But this signal is not where you want it to be (it is probably not even in a ham radio band), and the power level is too low even by the standards of the most fanatical members of the G-QRP club. So a few additional stages are needed.

To get the signal into the desired ham band, and make it moveable within that band, it is put through another mixer, a mixer that is connected to a variable frequency oscillator. In the case of my rig, I would have the SSB signal at 5.173 MHz and I wanted it to be in the 18.130 range. So I needed to mix it with a signal of around 12.9 MHz. I would make that 12.9 MHz oscillator variable in frequency, and this would be the circuit that would be attached to the big "FREQUENCY" knob on the front panel. After this mixer, RF amplifiers would boost the signal to the desired power level.

Geographic isolation surely contributed to the way this new transmitter took form. While mail order was available, being nine hundred miles out in the Atlantic made me more reliant than usual on my vast collection of junk box parts.

I'd learned to start a new rig by envisioning what it would look like at the end of the project—early on, I'd give some thought to the kind of box would I eventually put it in. I had an old Heathkit DX-40 that I had picked up at a Maryland hamfest. This rig was seriously defunct when I bought it, and I had increased its defunct-ness by cannibalizing it for parts. It was by now a shell of its former self. Fans of old radios don't like actions that result in the destruction of beloved boatanchors—this is viewed as an attack on our electromagnetic cultural heritage—so I was troubled by my plans for a new round of Heath-icide. I had prior offenses in this area: Remember how I turned that Benton Harbor Lunchbox into a mate for the Mighty Midget? But, hey, I was 900 miles out, in the middle of the ocean. And that rig hardly even looked like a Heathkit anymore. Someone had spray painted it silver! These are the kinds of rationalizations that one uses in this situation. Soon I was removing from the chassis all of the old parts that remained.

As in many of my projects, my old friend Pericles (RIP) helped out with this one. The same old lightning-fried Swan 240 that had provided parts for the Barebones Superhet and Mighty Midget projects coughed up key parts for this new endeavor. When I'd stripped that old rig down many years before, I'd wrapped up and carefully stored the four crystals that had formed the filter, and the crystals that were used in the main oscillator. Crystals and filters can be one of the major expenses in an SSB project—the parts from Pericles let me avoid this expense. And while modern crystal filters come as one manufactured component—one little mysterious black box—by using the four individual Swan crystals this project became even more completely homebrew. The markings indicated that these rocks had been made in 1962. Even at this point I could think ahead and imagine how cool it would be to lean back in my chair and tell some distant operator, "Well, the filter in this rig came out of an old Swan 240… crystals are of 1962 vintage." Wizard status was all but guaranteed.

I built this rig during the winter of 2001-2002. As with the double-sideband project, I used the Manhattan construction technique. I'd become an adherent of the isolation pad and crazy glue method. Troubleshooting and modification were a lot easier with all the components and connections on the top side of the circuit board. And I no longer had to walk around with green stains on my fingers and clothes from the chemical etchant (but now my fingers were usually covered with hardened superglue.)

The boards for the various stages started to fill up and come into operation: Crystal controlled carrier oscillator, audio (mic) amplifier, balanced modulator, Swan 240 crystal filter, RF amplifiers. Each stage was tested, and then I started the process of getting them to work together.

I'm sure that real professional engineers would shake their heads at what I was doing. The method I was using was very amateurish. I was borrowing bits of circuit design from many sources, and gathering components from several different generations of electronic hardware. I was putting a 1962 crystal filter from a vacuum tube radio right alongside an integrated circuit chip (the NE602) designed for use in cellular telephones. This is not the way professionals would design and build a new device, and this odd mixture of old and new technologies would never be done in the professional realm. But I am an amateur, and amateurs have a license to hack.

Going into this project, I thought the hard part would be the sideband generation unit—the intersection of balanced modulator and crystal filter. Surprisingly, that all went together fairly easily (at least at first—problems in this area did crop up later).

The carrier oscillator seemed to be working well—with my receiver I could hear its very loud and clear signal on 5.173 MHz. I built an audio amplifier using an LM386 integrated circuit chip. I prefer not to use chips, but it does seem easier to get audio amplifiers stable when they are built around a single chip (versus a number of discrete transistors). My ancient oscilloscope indicated that it was indeed amplifying audio. The balanced modulator was a repeat of the circuit that I'd used in the double sideband rig, so I was well familiar with the indications of success in this stage.

Then the filter was added. This too was relatively easy to test. An audio signal sent to the balanced modulator through the LM386 chip would produce the familiar double sideband signal, but at the output of the filter I could tell that only one of the sidebands was being passed. In fact, even without amplification, if I attached a microphone to the device and spoke into it while listening on a nearby receiver, I could hear my signal, and by tuning around as I spoke I could tell that the desired filtering action was taking place.

But so far I had built an extremely low power, fixed frequency transmitter. And the frequency that it was fixed on wasn't even in an amateur radio frequency band. It was now time to add the other mixer, and the variable oscillator that would allow me to move around within the frequency band.

Perhaps because I was anxious to get this rig on the air, I again deviated from my rule against using mysterious black-box IC chips. For this second mixer I chose an NE602 IC (the cell phone chip). In a tiny package the size of a fingernail there is an amplifier, a doubly balanced mixer, and an oscillator. Put two different signals into this device, and at the output you will see only the sum and difference products of the mixing—the original two signals will have been balanced out.

I could have built my own mixer circuit using diodes, but this would have been more complicated. These diode-based mixers require quite a bit of power from the local oscillator—you end up having to build one or two amplifier stages between your local oscillator and the mixer. You can drive the NE602 directly with the output of a simple one-transistor oscillator. This is a good example of how ICs make circuitry simpler.

At this point, I rather foolishly thought that I was home free. Disregarding recent experiences with the DSB rig, I thought that adding the needed RF amplification would be easy.

My first problem resulted from an effort to feed the output of the second mixer into the RF amplifiers. The NE602 had balanced out both the 5.173 MHz and the 12.9 MHz input signals, but I was overlooking the fact that it was outputting both the sum and difference of these two signals. So, in addition to the desired signal at 18.1 MHz, I was sending to the RF amps a robust signal at around 7.73 MHz. Ah, so that's why all the circuit designs always have another filter following the second mixer. Using coils and capacitors, I built a filter that would allow the 18.1 MHz signal through while knocking down the undesired 7.73 MHz signal.

The design and construction of this simple filter was very educational. It gave me some practical exposure to the concept of "loaded Q." Q is the quality factor for a tuned circuit. It is essentially a measure of the sharpness of the filter. It is largely determined by the amount of ordinary resistance in the coil. To put it in very un-scientific terms, the more resistance there is in a tuned circuit, the less dramatic will be the sudden disappearance of reactance at the resonant frequency point. The frequency response curve of the filter will be flatter and blunter, a rolling hill, not an Alpine peak. The term "loaded Q" recognizes the fact that the resistances in the circuit to which the filter is attached also affect its Q. I found that putting a bipolar junction transistor with low input impedance after the filter dramatically lowered the Q—it became more difficult to tune the filter for a sharp peak at 18.1 MHz, and it became more difficult to tune out the undesired 7.73 MHz signals. When I switched to a field effect transistor circuit with high input impedance, we were back to Alpine peaks, and the 7.73 MHz signals were left in the deep valleys of the response curve.

But my problems were far from over. Homebrew guru C.F. Rockey, W9SCH, put it this way, "It is the development of RF power in appreciable quantities that is the greatest challenge in this field for the average amateur." Right you are Rockey! I had a surprising amount of trouble boosting the carefully tailored SSB signal that was coming out of the Swan filter. Rockey went on to describe the mysterious forces that work against the struggling amateur. "Quantum mechanical necromancy" (the tendency of transistors to suddenly and inexplicably die) was among them. Rockey suggested that in extreme cases, the radio amateur might have to delve into the occult, and make sacrifices to the gods (he indicated that Papa Legba might be appeased by slaughtering a few chickens).

QST ran an article on this project, and at about this point in the story, their editor inserted the subheading, "NO CHICKENS DIED IN THE MAKING OF THIS RADIO." That's true, but that's only because no chickens were readily available. I had real trouble with those RF amplifiers. They would suddenly turn into oscillators. Or the transistors would burn up (often scorching my fingers in the process). Or they would just refuse to amplify. This was very frustrating, because I knew that I had an SSB signal in there, just waiting to be launched out to the ionosphere.

I think I was bumping up against the limits of hacking. As your projects grow more complex, there comes a point where better design skills can save you from a lot of cut-and-try agony. I was clearly at that point. I simply didn't know enough about amplifier design to understand what was going wrong.

I must have built and destroyed two or three final amplifiers before I threw my hands up in frustration. I was reminded of bad days in my ham radio youth, days in which I sat there in front of equipment that refused to work and asked myself why I had chosen such a frustrating and painful pastime. Finally, I decided to do what I had done with the double sideband rig: raid a previous project. The final amplifier from my 30 meter CW transmitter (the Santo Domingo project) was already sitting in the 17 meter DSB rig. I didn't dare butcher that rig—it was my only connection to my friends on 17. But my original 20 meter transmitter was still sitting there, ripe for slaughter. It had survived the first round of carnage because I had a sentimental connection to that rig. It had been my first real homebrew success. Chopping up the 30 meter transmitter had been relatively easy, but I didn't really want to cannibalize the 20 meter rig. I fought off the urge for a few days, and tried a couple more times to get a new amplifier going, but it was no use. In frustration I turned to my beloved VXO-controlled 6 watter. And I ripped its heart out.

As I had done with the DSB rig, I modified this amplifier so that it could be used with voice signals. It had been designed as a Class C amplifier, but with the addition of a few parts I was able to put approximately .6 volts of bias on the base of the transistors. This turned it into a much more linear Class AB amp.

And it worked. I could see that I was now getting about 5 watts out to the antenna. Best of all, the 5 watts were actually on the desired frequency, were not accompanied by smoke or flames, and appeared to be single sideband.

One of my early contacts with the new rig was Ed, WB6KOK, in Maine. Ed had been a big fan of my unusual double sideband signals, and would always delight in switching back and forth from lower to upper sideband while listening to me—he would note no difference when doing so. But now I was on the air with only one sideband, and Ed didn't know that I'd changed rigs. He heard me calling CQ and responded. During my CQ call, he'd hit the lower sideband button on his transceiver… and was astonished to find that half my signal had disappeared. "Hey Bill, something has happened to your signal! I just tried to listen to your lower sideband, and it is not there!" That was exactly what I wanted to hear. Thanks Ed!

Now it was time for some debugging and fine tuning. I needed to make sure that the frequency of the carrier oscillator was in the right spot relative to the passband of the crystal filter. If it was set too high, the filter would be chopping off high notes in my voice that were needed for communications clarity, and it would allow too much of what remained of the carrier (residuals from the balance modulator) through. If it was set too low, the voice signal transmitted would be lacking needed base notes. I didn't have the test gear needed to perform this adjustment properly, but my friend Rolf, SM4FQW, up in Sweden came to my aid.

One night, during a conversation with Rolf, I explained my problem and he offered to help me make the adjustments… by ear. Performing an electronic version of open-heart surgery, with power on and Rolf on frequency, I opened the case of the new transmitter. The carrier oscillator has a small capacitor that allows the frequency of the crystal to be moved slightly. With Rolf listening carefully, I would take my screwdriver and give that little capacitor a quarter turn to the right. "Better or worse?" I would ask.

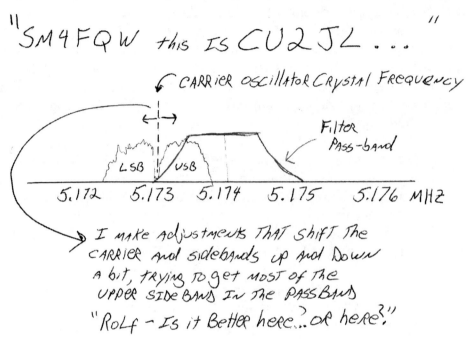

I think this little adjustment session captures much of the allure of ham radio. There I was, out in the North Atlantic, late at night hunched over a transmitter that had been forged from old Swans and Heathkits, from cell phone chips, and from bits of design from distant members of the fraternity of solder smoke. Pericles, the source of many of the key parts, was gone. So was Frank Lee, the amateur whose *SPRAT* article had inspired the project. But Rolf and I carried on with the core tradition of the radio fraternity: hams help their fellow hams overcome technical difficulties.

The North Atlantic is, I think, an especially good place for radio work. It is an angry, difficult, and dangerous ocean—it is a place where radio communication seems especially useful. This was the ocean that was first spanned by radio (Marconi 1901) and it was on this ocean that the very first voices carried by radio were heard (Fessenden's transmission to ship operators, Christmas Eve, 1906). With that dark angry sea all around me, linked only by a few watts of power reflecting off a faltering ionosphere, Rolf, a friend I had never met, helped me get that carrier oscillator exactly where it needed to be.

There were, of course, still demons to be exorcised. One of the most vexing came as the result of my effort to increase the frequency coverage of the rig. The 12.9 MHz crystals in the oscillator that fed the second mixer could only be "wiggled" so far. Maybe 9 or 10 kHz. Even with two crystals, I could only operate on 20 KHz of a band that is about 50 KHz wide. This caused a lot of frustration—a friend would be calling CQ just outside my transmitter's tuning range—I could hear him perfectly, but the limited range of those crystals prevented me from putting my transmitter on his frequency.

I knew that the higher the frequency of the crystal, the more its frequency could be shifted. With that 5.173 MHz SSB signal coming out of the old Swan filter, remembering that mixers produce both sum and difference frequencies, I realized that instead of having the local oscillator at 12.9 MHz and using the *sum* product coming out of the mixer, I could also set it up so that the oscillator would be running at around 23.3 MHz—under this scheme I would make use of the *difference* output of the mixer. With these higher frequency crystals I expected to be able to get about 23 KHz of shift. Two carefully chosen rocks would, I hoped, cover the entire band. Gone would be the frustrations of limited frequency mobility. With aspirations of greatly increased tuning range, I sent in an order for new crystals.

After a few weeks, the package arrived. I popped in the new crystals and made some adjustments to the oscillator. With the first of the new crystals everything seemed to be working as expected—I was getting the hoped-for improvement in frequency coverage. But when I threw the front panel switch and went to the second crystal, I knew that I had problems.

The vast majority of today's hams use commercial transceivers. Most have never used a station in which the transmitter and receiver were completely separate devices. But for most of ham radio's history, this kind of separation was the norm. You built or bought a receiver, and then later you built or bought your transmitter. Today, almost everyone is using a transceiver that automatically sets the transmitter on the receive frequency, but operators using the older, separate arrangement must manually place their transmitter on the frequency of the receiver. This was where my trouble developed.

Usually this task of "netting" (putting the transmitter on the receiver's frequency) can be done by ear. The receiver is left in operation. A low power signal from the transmitter is generated. Then the transmitter's frequency dial is turned until the signal is heard in the receiver. When the transmitter is exactly on the receive frequency, the tone coming out of the receiver will disappear. Recall from our discussion of mixers that these circuits produce sum and difference outputs. In a receiver the difference between the BFO frequency and the incoming signal from the crystal filter produces the audio that we listen to. If the signal from the crystal filter is at exactly the frequency of the BFO, no audio—no beat tone—is produced. This is known as "zero beat" and it indicates that the receiver and transmitter are on the same frequency, "netted," and ready to be used in two-way communication.

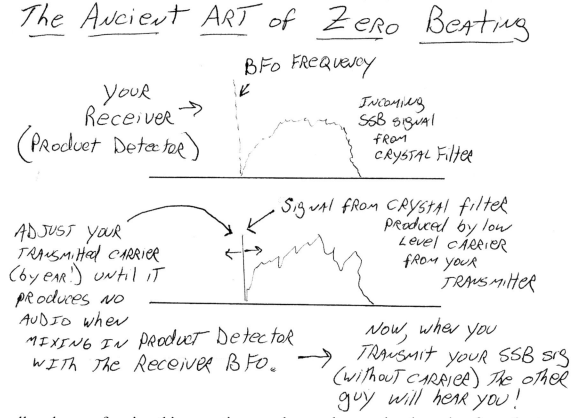

Normally when performing this operation you hear only one signal coming from the transmitter. But when I switched to the second crystal and tried to zero beat, I could hear multiple signals coming through. I could zero beat any number of signals. Clearly this was not good. Not only would I be unable to know if the signal I selected actually represented my transmit frequency, it seemed that with this second crystal my rig was transmitting on numerous frequencies at the same time. Having built this rig because I wanted to be a good radio citizen and conserve spectrum, this was a particularly galling problem.

My first thought was that there was something wrong with the crystal. This notion was supported by the fact that the other crystal seemed to work just fine. So I sent the rock back to the manufacturer (JAN Crystals) and asked them to check it out. They did, and they told me that it was fine. Damn! Now I had to troubleshoot.

I turned to the internet and got some good suggestions. But try as I might, I just couldn't get rid of those strange "extra signals." I couldn't even figure out why they were there.

After a lot of haphazard "lets-see-if-this-works" attempts at solving the problem, I finally decided to just sit down with a notebook and try to figure out what was going wrong. I noticed that the problem seemed to intensify as I tuned the transmitter close to one particular frequency: 18.116 MHz. I drew up charts of all the harmonics that were being generated by both the fixed carrier oscillator and the variable oscillator. Suddenly I saw the problem: As I tuned to that one particular frequency, strong harmonics coming out of the 5.173 MHz carrier oscillator started mixing with the output of the variable oscillator. These frequencies were being combined in the mixer and were generating additional signals that were making it through the 5.174 MHz filter—these were the additional signals that were driving me nuts.

At first I thought the solution was shielding. I thought I'd have to put most of the rig into separate, RF-proof enclosures. I was not happy with this prospect. Metal work and cabinet making are among my least favorite parts of radio construction. So before I launched into a shielding campaign, I gave the problem some more thought. It occurred to me that if I could just reduce the amount of harmonic energy being generated by that carrier oscillator, I might be able to eliminate the problem without having to build half a dozen RF-tight boxes.

A look at the schematic of Frank Lee's oscillator seemed to reveal a solution. It was a Colpitts design and used a capacitive divider to take energy from the output circuit and feed it back to the input. I realized that by varying the amount of capacitance from the base to ground, I could vary the amount of feedback and control the amount of harmonic energy being generated.

This solution came to me on an evening in which my official responsibilities required my wife and me to attend a reception. I'm not a big fan of diplomatic receptions, but this one was a good one. It was on the Portuguese Navy's beautiful tall-masted sailing ship, the *Sagres,* which at that point was sitting in Ponta Delgada's harbor. I'd been kind of looking forward to this event, but I knew that unless I gave this oscillator solution a try, I'd be thinking about my radio problems all night and would be unable to enjoy the reception. I hurried, and dressed as fast as I could. As my wife continued to get ready, I headed down to the shack.

The old issues of *QST* sometimes show the gentlemen hams of yesteryear dressed in suit and tie while working with their inventions. At CU2JL, we were normally much less formal. This was the first time I had ever had to worry about my necktie coming into contact with my soldering iron.

I had to work fast. The final touches of lipstick and perfume were being applied upstairs. Our driver had the engine running, and I knew that many of my Consular colleagues were at that moment heading for the gangplank of the *Sagres*. But I had a trimmer capacitor to install. I had harmonics to squash.

I opened the cabinet and soldered in place a ceramic trimmer capacitor. Then, as the high-heeled shoes clicked down the stairs, I tuned my workbench receiver to the hated spurious signals that had been haunting me for weeks. With a screwdriver, I tightened the ceramic capacitor's screw, increasing the amount of capacitance between the base of the transistor and ground, thus decreasing the amount of feedback in the oscillator. As I turned the screw, the gremlin signals rapidly faded out of existence. It was like an electronic exorcism. It was as if I had squeezed the life out of those nasty beasts. A quick check revealed that the desired frequency was still there, and output from the transmitter was still very good.

The shack door swung open. "Ready to go honey?" asked my wife. With a smile on my face and a whiff of solder smoke in my hair, we drove off. It was a very nice evening. Of course, I told no one other than Elisa and Mr. Silva of my technical triumph.

This had been an enormously satisfying debugging session—I had to use my understanding of electronic theory to solve a unique problem in a unique rig. Good test gear helps a lot, and the advice that is available from friends on the internet is very useful, but sometimes you just have to sit down with the schematic and a notebook and try to figure out the likely cause of the problem at hand.

In his wonderful book, *Surely You're Joking Mr. Feynman – Adventures of a Curious Character*, Richard Feynman describes his adventures as a Depression-era radio repairman in Brooklyn. Feynman was still a boy when he got into the radio repair game. On one job, the owner of the broken radio complained when young Richard seemed to waste time by pacing up and down in front of the device. The owner didn't realize that he was engaged in the most important part of the repair process— thinking the problem through. After some more pacing, Feynman reached into the cabinet and quickly made the needed repair. The owner was amazed, and told everyone about Feynman's astonishing ability: "He fixes radios by thinking!" Indeed.

From time to time it is good to take a break from the workbench and engage in other pastimes. Fortunately for me, I had available a seemingly endless supply. I understand there are people who have trouble keeping themselves busy in retirement. I do not think I will have that problem.

For example, there's rocketry. This had been one of my passions as a kid. When we were back in Santo Domingo, during a visit to the local hobby shop Elisa had noticed that I was all but drooling over one of the Estes model rocket kits. It was the Astrocam, the only rocket with an actual camera in the nose cone. With the exception of the Astron X-Ray, the Astrocam was the only rocket in the Estes fleet that did anything useful payload-wise (the X-ray had a large transparent cargo chamber that was used to test the effects of G-forces on a wide variety of unfortunate small animals).

John Carlson had been the only member of the Waters Edge Rocket Research Society (WERRS) to build and fly the Astrocam. At the time, it had seemed far beyond my technical and financial capabilities. Elisa noticed my interest, and when Christmas rolled around, a genuine Estes Astrocam rocket kit was under the tree. I was so pleased with this present—I felt like I was 12 years-old again. On some of the rocketry web sites they talk about "Born Again Rocketeers"—guys who built and flew rockets as kids, and then resume the activity in middle age. That was me.

We flew the rocket several times in Santo Domingo and got some great pictures. It was then in storage until the Azores. On Sao Miguel Island there was a very active model airplane group. I linked up with them, and as they flew their planes, I'd launch the rocket. My goal was to get a good aerial shot of their landing strip. Of course, the rocket launches contributed further to the by-now widespread suspicions about "O Consul's" mental equilibrium. But I'll bet people murmured about Werner Von Braun and Robert Goddard too. This is the price you pay for being a man of science.

The Astrocam project led me to another aerial photography project, this one potentially even more damaging to my reputation. A rocket is not the easiest of platforms for photography. I had a difficult time getting the desired picture of the landing strip. There must be a better way, I thought… Of course! Kites! Put the camera on a kite and you'd have a much more stable platform and much better control over your shots.

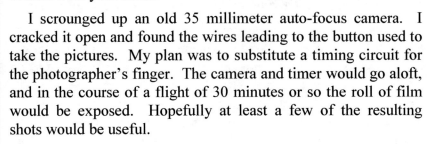

I scrounged up an old 35 millimeter auto-focus camera. I cracked it open and found the wires leading to the button used to take the pictures. My plan was to substitute a timing circuit for the photographer's finger. The camera and timer would go aloft, and in the course of a flight of 30 minutes or so the roll of film would be exposed. Hopefully at least a few of the resulting shots would be useful.

The NE555 chip was an obvious choice for the timer circuitry. Together with a few resistors and capacitors, this chip could do the needed high-altitude button pushing. But I ran into a bit of a problem. It turned out that my camera needed the button to be held down for about a half second. So I had to design some additional circuitry to keep the firing contacts closed for the needed interval.

Next my thoughts turned to the kite. *SPRAT* had an article about a kite designed to serve as a support for wire antennas. It is a "Scott Sled" design. I figured that it would have sufficient lift to carry my camera and timer aloft. Just to be on the safe side I scaled it up by about 50 percent. I built it with black plastic garbage bags, Duct Tape, and—for the vertical supports—pieces from the same batch of fishing poles that I used in the antenna project.

Now I had to figure out how to suspend the camera and timer under the kite. I turned to the net. I thought it would take a lot of Googling to find information about camera mounts for kites, but I was surprised to find that I was far from the only one interested in this kind of thing. There were hundreds of Kite Aerial Photography pages. I was not the only one who had thought of linking camera with kite. The web sites provided construction details on a number of camera mount schemes. I went with a simple pendulum mount—the camera would hang from the line. A one foot length of wood (from a yardstick) would replace a section of the kite string. From this piece of wood, a second length of wood would pivot down like the swinging arm of a pendulum. Gravity would keep this arm vertical. The camera and timer would be mounted on the swinging arm.

This has to be the strangest kite ever to have flown over Sao Miguel Island. It was an ominous looking beast. Most kites are made in bright, happy colors. This thing was garbage bag black. The Scot sled design had two big vent holes in it—they looked kind of like evil eyes staring down. I put a long orange tail on the thing to help prevent it from looping and swooping. The strangest feature was, of course, the photographic payload that bobbed in the wind a few feet below the kite.

Flying this thing was hazardous, both to the population and my reputation. The winds were usually on-shore, so the big kite and the camera would be suspended over the beach shops and the coastal road. Every once in a while my American fear of lawsuits would kick in and tell me to haul this thing back to earth before some innocent Azorean got conked on the *cabeza* by my airborne 35 millimeter, but then I would just look around the beach and note the pit bulls romping happily (leash free) through the dunes, the trail-bikes zipping along the beach, and the lawnmower-powered paraglider that occasionally shared the skies with my contraption. My kite seemed like a minor risk compared to other Azorean beach activities. Still, the possible headlines were appalling: "AZOREAN GRANDMOTHER KILLED BY AMERICAN CONSUL'S MONSTROUS CAMERA-KITE!" But hey, no guts no glory. Up the kite went.

And it worked! The first time out we got some very good pictures of the beach. We could see yours truly standing in the sand with the string leading up to the kite, Elisa and the kids wisely huddling for cover under our beach umbrella. We tried several times to get a good shot of the model airplane airport, but each time we tried, the winds failed us. (This was really amazing and frustrating, because it seemed like the winds had been blowing incessantly for the previous three years.) But we did get the beach shots. The project was a success. And no Azoreans were killed in the making of those pictures.

The winds that held the kite camera aloft soon started to form into the familiar tornado that we knew would sweep us out of the Azores. We knew this would be an especially difficult departure.

I wanted to give something to Messias, so during our last months in Ponta Delgada I decided to build for him a special homebrew transmitter. I scrounged around in the junkbox and found a circuit board for a very simple crystal controlled 80 meter Morse code transmitter. Crystal control would be tedious, and I knew that it would be easy to make a real variable frequency oscillator for those frequencies. After a bit of design work and some experimentation, I had one percolating along. I built the transmitter on a nice varnished piece of hardwood and left all the circuitry exposed. It looked like a transistorized version of the homebrew tube type transmitters of yesteryear. Stan, Bollis, and Serge would have liked it.

Messias was almost moved to tears. On the day I gave it to him we met up on 80 meters and actually used the thing to talk across Ponta Delgada. A few days later he reported that he had used the transmitter to contact an impressive list of far distant stations. I'm not sure about those contacts—he may have been trying to make me feel good. After we left the islands, the local TV station visited Messias's shack and talked to him about ham radio. He pulled out the homebrew transmitter and told the story of how it was built.

It seems like my wife and I spent those last few weeks in the Azores constantly on the verge of tears. (I was usually on the verge, she was usually over the edge.) The Azoreans are very serious about friendship, and they are not all that used to having friends move away. So just about every encounter during those final days drew a sad comment along the lines of, "Well, I guess this is the last time we'll be bumping into you here in the market," followed by tears and hugs.

Finally, D-day arrived and off we went. It was a direct transfer—no stop over in Washington, no trip back home. Breakfast in Ponta Delgada, lunch at the airport in Lisbon, and dinner in....

CHAPTER 6
URBAN RADIO
SOLDER SMOKE IN CENTRAL LONDON

"In our school years, Hal and I filled most of the third floor with working ham-radio transmitters and receivers. Our rigs were mostly built from a mixture of post-war surplus equipment and junk television sets. We learned by experience that when you need high voltage, the power company's 6,000-to-120-volt transformers work admirably in reverse; and that most amplifiers will oscillate, especially if you don't want them to."

—Joe Taylor, radio amateur (K1JT) and winner of the Nobel Prize for Physics in 1993, describing the boyhood shack that he shared with his brother.

I'd finally gotten my homebrew 40 meter double sideband (DSB) transmitter working and I was anxious to try it out. This was the first rig that I'd built in the UK. As I prepared to make my first calls, I tried to think of an appropriate name for the transmitter. "The Anglo-American"? Maybe "The Yank"? Perhaps "The Churchill"?

Unfortunately on the day of that first on-the-air test conditions were poor. My Central London location imposed severe handicaps in the antenna area, and I was putting out only a few watts of DSB. So I knew that I wouldn't be threatening the health of anyone's receiver. But as the Brits say, I thought I'd "give it a go."

After much searching for a clear frequency (you have to be doubly careful with DSB) and much futile CQ-ing with my newly minted UK call sign (M0HBR—ambitiously: HomeBrewRadio), I finally made contact.

It was a British station. I was delighted, because it was precisely for UK contacts that I'd built this rig. As a teenage ham in New York, contacts with the UK were among my favorites. Crossing the pond was always great, but for some reason there was something special about the contact when a G-station was at the other end. Much later, in the Azores, I was almost exactly one ionospheric F-layer hop from the UK—it seemed like at least a third of my contacts were with UK amateurs. When I moved to London (and inside the UK skip zone) contact with UK became nearly impossible on 17 and 20 meters, and I quickly came to miss my conversations with our congenial British cousins. So, with local contacts in mind, I built a 40 meter rig. I was hoping for Near Vertical Incidence Skywave (NVIS) contacts: my radio waves would still be bouncing off the ionosphere, but they would be going pretty much straight up and then straight back down.

On this first day of tests, the other station wasn't hearing me very well, and the band was starting to fade, but I couldn't resist asking him for an audio report.

"Well Old Man," he said, "I think you may have some sort of problem in the audio amplifier... There seems to be some distortion... I can't quite put my finger on it, but it somehow is making you sound... How can I put it? IT IS MAKING YOU SOUND LIKE AN AMERICAN!"

With a chuckle, I politely explained that the distortion he was hearing had been injected into the system some 45 years earlier in New York City, and that tweaking the audio amplifiers was not going to get rid of it…

That Foreign Service tornado had struck again, lifting us out of the Azores and plunking us down on another, larger island some 1500 miles to the Northeast. We'd arrived in England in August 2000 directly from Portugal. It was breakfast in Ponta Delgada, lunch in the Lisbon airport, and dinner in our new home in London's South Kensington neighborhood.

It was quite a change. In one day we'd moved from the middle of nowhere to the middle of everything. South Kensington (South Ken to the residents) is part of what is formally known as "The Royal Borough of Kensington and Chelsea," and the place lives up to its rather pretentious name. Almost immediately we started bumping into movie stars and other celebrities. Madonna was reportedly in the 'hood. The actor Hugh Grant was a regular at our newspaper kiosk. The Sultan of Brunei was said to have a little place around the corner.

Of course, I immediately started a radio amateur's evaluation of the new location. The house had a small upstairs room that looked like it could be pressed into service as a shack. There was no back yard to speak of, but I did have access to the roof, so there were some antenna options available. But the small house was at the bottom of a brick canyon, with large apartment houses towering around it—getting a signal out would be tough. My gear had not yet arrived, but I had in my hand luggage a small shortwave receiver. On that first day in the house I tuned around and was relieved to find that the RF noise level (often a problem in urban locations) was not too bad.

As we moved around the new neighborhood in those first few weeks I continued to size the place up from an electromagnetic standpoint. There were bad signs. I had looked forward to easier access to electronic parts, but I soon discovered that in South Ken, you couldn't find an electrolytic capacitor even if your life depended on it. There were miles and miles of antique shops, boutiques of every conceivable variety, stores selling every known variety of absurdly expensive handbag and perfume, but there was no place to buy a 1 kilo ohm resistor. There were crystals on sale, but only of the opium-based or New Age varieties. There were Starbucks and real estate agents and art galleries on every block, but in all of South Ken there was not a single purveyor of electronic components, not even a single Radio Shack store, or even the rather pale UK equivalent, Maplin's. I was shocked to find that I had easier access to electronic components in Ponta Delgada than I did in London.

London shack

A scan of the rooftops confirmed my suspicion that I was in an area devoid of amateur radio; there were no amateur radio antennas up there, not even a little dipole or a long wire poking out of an attic window to indicate that a fellow fanatic was melting solder or seeking signals from beyond the seas. Occasionally, I'd spot a large beam antenna and my hopes would be raised, but as we got closer I'd see that the array was hovering over some embassy, and was aimed permanently at the Ministry of Foreign Affairs of some difficult-to-pronounce foreign capital. From time-to-time I'd spot a car license plate that looked like an amateur radio call sign (in the States you can have your call sign on your license plate) but after a while I realized that these were just random accidents.

Even the bookstores provided evidence that I had entered a radio-free zone. There'd be shelf after shelf devoted to the arts, biographies (of artists!), interior decorating, etc. There would usually be a section called "popular science," but nothing on radio or technology. Not even computer books. Clearly, I was living in the arts and humanities section of London. I feared that I had moved into an area completely free of the kind of radio-activity that I was looking for.

This was somewhat ironic and surprising, because we were also living almost on the campus of London's Imperial College, one of several universities described as "the British MIT." There is a Yagi beam antenna atop one of Imperial's buildings, but it is sagging and twisted and obviously not in use. I Googled and Googled in search of an Imperial College Amateur Radio Club. No joy.

As far as I could tell, other than my little radio shack, the only pocket of amateur radio in the Royal Borough was in the nearby Science Museum, where there was a nice display of radio technology from days-gone-by.

I was disappointed by all this, but I decided to soldier on. I guess I really decided to solder on. I would make it my mission to keep the Royal Borough active on the amateur radio frequencies, and I would melt some solder in South Ken. That 40 meter DSB rig was my first London project.

Some homebrew rigs come together very easily. Others do not. The supposedly simple double sideband rig fell into the latter category.

First I had trouble with the oscillator (it's a bad omen when your troubles begin this early). I have always used variable crystal oscillators, but up until this point most of my rigs had been for 14 or 18 MHz. At these higher frequencies it is possible to get useable amounts of frequency shift from a quartz crystal, but as you go down in frequency, the amount of available "wiggle" decreases. I did not relish the prospect spending a good bit of money on a crystal only to find myself restricted to a tuning range of 5 kHz. This would be adding insult to injury: I knew I was already facing the handicaps of a poor antenna in the urban canyons of central London, QRP power levels, and half my RF energy being devoted to an unheard sideband. A 5 kHz tuning range (and an expensive one to boot) seemed a bit too much.

I decided to get daring. I was going to go way beyond my usual crystal wiggling and build something vaguely synthesizer-like.

I suppose calling it a synthesizer is a bit of a stretch, but it makes me feel better about the failure that ensued. (It's easier to accept failure when the project is complicated and difficult) I figured it would be easy to build a stable VFO in the 3 MHz range, and I had plenty of crystals in the 10 MHz range. A simple mixer should have me gliding effortlessly across the length and breadth of 40 meters. In planning a homebrew project, anytime you find yourself using words like "simple" or "effortlessly" or "easily" alarm bells should go off. Loud alarm bells.

The 3 MHz oscillator went together easily enough, and I learned a lot about how to get the circuit to oscillate in the desired range. There were no problems with the 10 MHz oscillator. The trouble came with the mixer.

I went through several versions. There was an early attempt at using a diode ring. Then I brought an old 40673 dual gate MOSFET out of retirement. But I kept getting a lot of 3 MHz and 10 MHz energy in the output. No good.

I then decided to use an NE602 chip. It's a double-balanced Gilbert Cell, so I thought it would be easy for me to keep the undesired 3 and 10 MHz energy out of the output. Thus began the slaughter of chips, the massacre of mixers.

Using a very old scheme, in an effort to improve stability, I'd decided to keep the 3 MHz variable frequency oscillator running all the time. The 10 MHz crystal oscillator and the mixer chip would be shut down when I went to receive. A *simple* little T/R relay would perform the switching magic.

The switching arrangement did work, but only for a while. Then the NE602 chip would die. I use a combination of "ugly" and "Manhattan" style construction. For most components, this makes replacement very easy. But with the eight pin IC's placed "dead bug" style on the board (no sockets) replacement is not so simple. I use very fine wire to connect the pins to pads glued to the PC board ground plane. Each one must be sort of lassoed and soldered. It isn't easy, especially with 47 year-old eyes. But I went through the exercise enough times to get good at it.

They say that madness is when you do the same thing over and over again, yet expect a different result each time. Well, I guess I was in the grip of homebrew madness, because for some reason I just kept blowing up NE602 chips, and replacing them with new ones. I guess I went through several theories of what was causing the failure: First I thought it was too much audio. Then I thought it was too much RF. Then I theorized that I might have too high an operating voltage. And on it went. I only stopped when I ran out of chips. An old American advertisement about potato chips came to mind: "You never can eat just one!"

What was the cause of this component carnage? Well, my latest theory is that high voltage transients from the relay coil were taking them out. My chips were being done in by the theories of Einstein, Maxwell, and Faraday. I would turn off the relay by simply cutting off the current flowing through the coil. But when I did this, there would be a massive, sudden change in the magnetic field around the coil. And a changing magnetic field creates an electric field. I think it was this electric field that was frying my chips. It's always a good idea to put a capacitor and diode across the coils on a relay—they will prevent the voltage that results from the changing magnetic field from damaging the circuitry.

At this point, I abandoned my dreams of synthesizers, and decided to see if I could build a more stable oscillator running at the 7 MHz operating frequency. Using a toroidal coil, NP0 fixed capacitors and a good quality variable capacitor, the oscillator went together very easily, and proved to be quite stable.

Traumatized by the synthesizer episode and completely out of NE602 mixers, I decided to use a simple two-diode single balanced mixer for the balanced modulator circuit. These circuits need a lot of RF drive, so I had to put some amplification between the oscillator and the input port of the diode mixer.

For my audio amplifier, I started out using two discrete junction transistors, but I had trouble getting it stable. So I went with the old stand-by: a 741 IC op amp circuit.

With RF and AF mixing together in my diode mixer, I thought it was time to move into the fascinating world of RF amplification.

This phase of the project is something of a blur now. A painful blur. You'd think that this would be the easy part. You'd think that getting the oscillator going and generating the DSB signal would be the tough part, and that the RF amplification bit would be relatively straight forward. Think again! By the time I was finished, the workbench looked like some sort of electronic battlefield. The remains of dead transistors and chips were all around me. A cloud of solder smoke hung in the air. And I myself was actually wounded: As a result of an heroic effort to use my fingers as test instruments ("Let's see if the amp is getting too hot"), the image of an RF power transistor had been burned, perhaps permanently, into my thumb. Scratch-built homebrewing is not for the faint-of-heart.

It was a ham radio flea market ("hamfest" in American, "rally" in British) at Kempton Park that provided the weapon that eventually won this battle: a nice old HAMEG oscilloscope. No longer was I operating in the blind. It was the workbench equivalent of night vision goggles, and it allowed me to see what was happening in the circuitry. I noticed that the signal coming out of the balanced modulator looked awful. I suspected harmonics and put a low pass filter between the balanced modulator and the PA. That took care of it.

During the testing phase of this project I came across another benefit of the simplicity of DSB rigs. When it comes time to test an SSB rig, you really need a two tone test generator. The two tones produce the familiar "sideways hourglass" oscilloscope test pattern that lets you look for amplifier flat-topping and other forms of distortion. But a DSB rig produces this pattern with only one audio input tone.

My 40 meter rig has a rustic look, the result of how and where I procured the chassis and cabinet. In a Dyas's hardware store in Windsor I found a kitchen cutting board that seemed just the right size for the chassis. At a hardware store in Chelsea I found an aluminum "kick panel" for a door that seemed just the right size to wrap around the cutting board—this would form the front panel and the sides of the cabinet. (The use of the kick panel was very appropriate, of course, because I often had the urge to kick the thing across the workshop.) Finally, the top of the rig's cabinet came to me off the streets of South Kensington: Walking to work one morning, I found a computer discarded by a neighbor who obviously had more money than technical ability. Of course, I took it home and stripped it bare. One of the side panels became the top of the 40 meter DSB rig's cabinet.

During my struggles with this transmitter, I posted some questions about it on the G-QRP e-mail mailing list. As always, I received many helpful replies, but I also got some quite anxious messages indicating that there was no place for DSB on 40 meters in the United Kingdom. The American call sign at the end of my inquiries seemed to have set off these anxious responses: It was as if I was proposing that we SUPER-SIZE all the signals, the radio equivalent of replacing London's beloved mini-Coopers with American SUVs or Hummers.

There's no doubt about it: DSB uses at least twice the amount of spectrum used by an SSB rig. But I think there is a place for this kind of rig on our bands. Surely in a hobby that has its roots in the homebrew radio adventures of technologically intrepid lads armed with soldering irons and ingenuity there must be some room among all those YeasuIcomKenwoods for a few homebrew DSB rigs at QRP power levels.

The rig went on the air. It regularly produced nice contacts with UK radio amateurs.

I still haven't decided on a name for it... Maybe "The Little Hummer?" Or "The South-Ken"?

Wait, I've got it! We'll call it "The Accent"!

For astronomy, the move to Central London wasn't a step in the right direction. I think the number of cloudy days is a bit lower than that of the Azores, but vastly increased light pollution seemed to balance out this slight advantage. Another factor was the portion of the sky that I could see—from our huge backyard in the Azores I had available to me perhaps 90 percent of the night sky. But from South Kensington in London I peered up through a canyon of brick from a postage stamp-sized garden area that couldn't really be called a yard. From this tiny garden, I could see perhaps fifteen percent of the night sky. Fortunately most of the view was to the south, so I could see the ecliptic—the path of the planets, the path of the only objects that were big and bright enough to study from South Kensington.

Jupiter's moons provided some fun. In January 2005 Jupiter was visible from the garden early in the morning. I'd consult a computer program to see where the moons were all located, then I'd go out with the telescope and a cup of coffee to watch them dance. As the sky brightened with dawn the four Galilean moons would disappear one by one. I noticed that Ganymede was the last one to disappear in the morning light… sure enough, the books revealed that this was the largest of the four. It was fun to make that discovery on my own.

And I was not the only one in the family making astronomical observations. On January 23, 2005 I took my four year-old daughter Maria to the Starbucks on Fulham Road. Maria had developed an expensive fondness for Chocolate Cream Frapucinos. On our way home, we checked to see if we could see any stars. "Look daddy, a spaceship!" South Kensington is in the approach pattern for Heathrow Airport, and a plane goes overhead about every 45 seconds, so my first thought was that she must be looking at a plane. But I looked at what she was pointing to and it had the very distinctive steady motion and steady light of a satellite. I noted the time: 1745 GMT. I told her I'd check the computer to see if she had in fact spotted a spaceship.

There is a wonderful website called "Heavens Above" that displays the orbits of many of the visible satellites. Considering the extreme light pollution we were dealing with, my guess was that the only satellite that could have been visible to us was the huge International Space Station. Sure enough, at 1745 that dark London afternoon, the Space Station was passing right above London. Not bad, Maria.

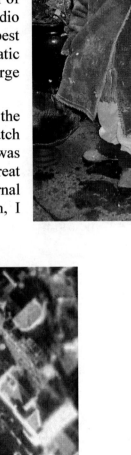

London is a very challenging environment not just for the reception of visible radiation from space, but also for the transmission of amateur radio signals. A tiny garden surrounded by concrete canyons is not the best place for erecting antennas. And then there are the legal/bureaucratic factors. I'm told that the local authorities are not crazy about large amateur radio antenna arrays atop Victorian-era houses.

Fortunately, I had access to the roof. In the hallway adjacent to the shack a small stairway (it was more like a permanent ladder) led to a hatch reminiscent of something you'd find on a submarine. Up above, I was immediately reminded that in addition to electrocution, the other great danger faced by radio amateurs is falling. Mindful of my paternal responsibilities, and concerned about the local planning commission, I decided to keep my antenna projects modest.

My antenna in London's concrete canyons.

I'd never put up antennas in an urban environment before. There were no trees at which I could hurl stones with ropes attached, and from which I could hang dipoles. Thankfully, service in London took away completely the notoriety we had lived with in the Azores. In London we were completely anonymous, and no one cared whether I was nuts or not. But the high population density and the close proximity at which people live can create some uncomfortable antenna situations. I'd be up there on the roof, hacking away at some bizarre-looking antenna when suddenly a neighbor in the apartment house adjacent to us would step out onto his balcony for a smoke and/or a cup of coffee; he'd be quite startled to find me in his field of view. To avoid these kinds of encounters I took to doing my antenna work very early in the morning.

I had brought with me from the Azores the fishing poles that I'd purchased in Ponta Delgada. The rotate-able dipole that had graced our Azorean roof had telescoped down to ship-able dimensions—it soon stretched its electromagnetic wings over the not-so-mean streets of South Ken.

If you are in London and had to point your finger (or your antenna) at the United States, which direction would you point? West is an obvious answer, but it is not quite correct. You have to remember that we are on a sphere and factor in the curvature of the earth. We are talking "Great Circle Paths," the tracks used by airliners. The shortest distances between A and B. On the globe an airplane (or a radio signal) going due West from London passes not through Iowa, but through Cuba.

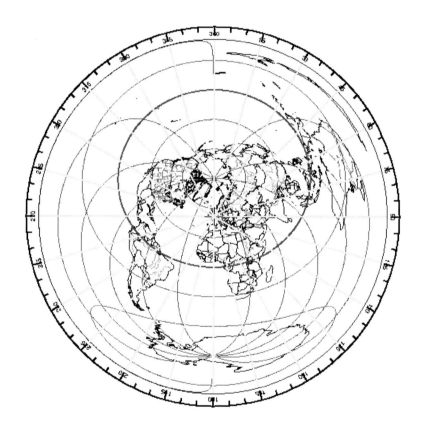

Great Circle Map from London

Azimuthal Equidistant Projection
From M0HBR, South Kensington
Radial scale: 2500km/cm
http://www.wm7d.net/azproj.shtml

AZ_PROJ v1.1.6beta5, Jan 2002, (C) 1994-2002 Joseph Mack NA3T, Michael Katzmann NV3Z

The American heartland was to the Northwest, and here the situation was hopeless, literally a brick wall. I knew from the start that I would not be having many conversations with hams in the homeland. There were also a lot of bricks and re-bar to the East. That made contacts in the direction of Germany and Poland quite difficult. The view to the South and Southwest was much better, so contacts with Spain were more frequent.

So there I was, trapped in a concrete canyon, armed only with simple, low-power homebrew ham gear. To make matters worse, by this time we were approaching the bottom of the solar cycle—sunspots were now very scarce, and the ionospheric layers that we used for long-distance communication often just faded away.

For a true radio amateur, there is no substitute for a real, on-the-air, shortwave radio contact. Some may wonder why we bother with all the radios and antennas, and all the uncertainty caused by sunspots and atmospheric conditions. Why not just pick up the phone and talk to whomever you want? Well, there is just something special about pulling a faint voice out of the ether using your own antenna and radio equipment, and then sending your voice back over the same ethereal path. The thrill is intensified greatly if you have built all or part of the equipment yourself.

For many of us, these amateur radio conversations become a significant part of our social lives. When those sunspots disappear, you find yourself missing the electromagnetic fraternity. It can get kind of lonely in the old ham shack. Fortunately, modern technology has provided a solution.

I mentioned earlier that Rolf and I had sometimes used a Voice-over-internet program called Echolink to stay in contact. Echolink is a system developed by Jonathan Taylor, K1RFD. It is a big improvement over the Internet Phone system that we were experimenting with in the late 90's. Gone are the porno fiends. No longer do radio amateurs have to hold their noses and prepare to cringe as they seek out internet-based contact with their fellow enthusiasts. For that alone, Jonathan deserves the thanks and admiration of amateurs around the world. But with his system he did much more.

Through Echolink the fiber optic cables of the internet now serve, in effect, as an artificial ionosphere. Instead of bouncing radio signals off the upper atmosphere, we are now bouncing digitized light signals through light pipes stretched along the floors of the oceans. It seems like a new frequency band, one that is always open and that works equally well for contacts across the street and across the globe. It is entirely free. Log onto Echolink (access is limited to licensed radio amateurs) and you will find a very user-friendly screen listing thousands of amateur stations that are also logged on. With a click of the mouse, you will be in voice contact with any of them. The audio quality will be excellent. There will be no interference, no static, no fading. Roundtables like those on the traditional ham bands are easily arranged.

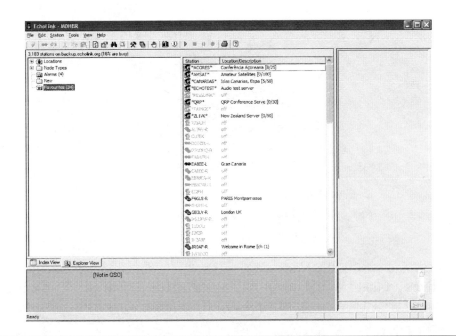

Perhaps most exciting are the possibilities of linking the internet with traditional amateur radio systems. I did a bit of this in the Dominican Republic when I connected I-phone to my old Yaesu Memorizer transceiver and took I-phone calls with my old walkie talkie. But Echolink allowed this to happen efficiently, and on a large scale. For example, with Echolink I can easily connect to the very high frequency (144 MHz) repeater of the radio club of Alice Springs in the Northern Territory of Australia. Without Echolink, this repeater covers only the 10-15 miles around Alice Springs. But Echolink allows me to reach out from the other side of the world, and call CQ on the Alice Springs repeater. Or on a similar repeater in Moscow, or Tokyo, or Hawaii.

It isn't "real radio" but the friends you can make on Echolink are definitely real, and the way these friendships are formed is identical to the way they are formed on the airwaves. There are now many other VOIP programs in use by the general public—you can talk to people in Australia for free on the Skype system, for example. But if you called up a randomly selected person and tried to strike up a conversation, I don't think you would get very far. Echolink, however, has clearly inherited the "CQ culture" of ham radio. On Echolink, you can randomly select someone, and give them a call. Just as on the ham bands, you'll never get a "What do you want and why are you calling me?" response. On Echolink, as on the amateur radio airwaves, they know why you are calling.

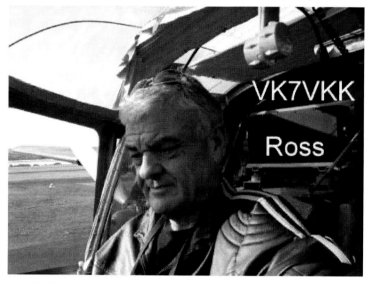

Ross, VK7VKK, lives on the Australian island of Tasmania. We met on Echolink and discovered a lot of common interests. In essence, we are both gadgeteers, tinkerers. Ross told me about his ultralight airplane, and his antennas. He'd set up VHF and HF links to the Echolink system—I could connect to his station and call CQ on the airwaves from Tasmania. Most of the time he'd be the one who answered—I'd catch him in his car on the way to the supermarket with his wife, Inge. I'd say hello to Inge and ask about the kids. Early on in our contacts Ross was building a new porch for his house—he'd give me updates on the project, send pictures of the progress, and occasionally vent some frustration with the whole thing. "How's Ross doing with the porch?" Elisa would ask. Months later, Ross took pleasure in answering my call on his walkie talkie while out on the porch, cooking dinner "on the barbee." Maintaining this kind of ordinary human contact with unmet friends from far away has, for me, always been one of the big attractions of ham radio. It adds an important element of warmth and society to a hobby that could easily be as cold and harsh as the modern computer culture.

I first came across this friendly, fraternal spirit in the stories of Jean Shepherd and in the 75 meter AM conversations that I monitored on my beloved Lafayette HA-600A (with jeweled movements). It was there in the Spanish language radio chatter on the local network of the Dominican Radio Club. And although the physical setting was literally out of this world, it was there in my conversations with Norm Thagard when he was on the MIR space station. It was the same mix of tech talk and "how'ya doin' today" conversation that has always been the hallmark of ham radio.

Over the years, I'd noticed that radio amateurs like to listen in to this kind of chatter. Even if you are not participating, if you find a congenial and interesting group of friends yakking away on the air, it makes for some very pleasant listening. Many guys will leave their receivers on as they build things on their workbenches. Believe me, it is far better than having the TV on. It always reminded me of one of the early tech-talk programs on public radio, "Car Talk." I'm not a car guy, but for some reason I liked listening to those two brothers from Boston talking about cars. Sometimes, out in the Azores, when Mike Higgins and I would end one of our long conversations on 17, other guys would call in saying how much they'd enjoyed listening in. That positive feedback got me thinking...

"The New Scientist" is one of the best science magazines in the world. The "Technology Trends" section in the April 16, 2005 issue caught my eye. Paraphrasing a line from an old rock song, there was an article entitled "Software Killed the Radio Star—Amateur DJs are abandoning the airwaves and going online to broadcast to the world's MP3 players". This was my first exposure to the idea of podcasting. For those of you who have been living in a cave, let me briefly explain. A podcaster records a program in much the same way as a traditional broadcast program (like "Car Talk") is prepared. But then, instead of sending the program to a network of syndicated AM or FM broadcast radio stations, the podcaster simply makes the recording available on a web site. Listeners can then listen to it on their computers, or download it to their portable music devices (mp3 players or iPods). The program can then be enjoyed at the listener's convenience.

The New Scientist article opened with the description of two podcasters who put together a twice weekly internet radio program. The unusual thing about this show was that one of the hosts was in California and the other was in the Netherlands. They used the Skype VOIP program to get together. This got me thinking about Echolink, and what we could do with it.

Mike Caughan, KL7R, was one of my Echolink friends. Mike lived in Juneau, Alaska. Since 2003 we'd been meeting on the system and discussing our homebrew radio projects. We were obviously kindred spirits: We were almost the same age, and in our first conversation Mike told me that he had converted an old Heathkit HW-8 CW transmitter into a Double Sideband rig. Soon we were talking weekly on Echolink, comparing notes on our projects. After reading the New Scientist article, I mentioned it to Mike. I noted that Echolink has a record feature....

Mike had a very go-get-'em, innovative spirit. And he had a lot more experience with software and computers than I did. A day or so after our talk about podcast possibilities, I started getting e-mails with audio file attachments from Mike. He'd taken the recording of one of our conversations and, using a bit of audio editing software called "Audacity" turned it into an mp3 file. He'd also come up with some intro music, and had found a website (Ourmedia) that would host the mp3 files for free. Thus was born the SolderSmoke podcast. We uploaded our first show on August 21, 2005.

So our podcast was really another one of those spontaneous combustion things that just seem to happen sometimes in a ham shack. But looking back, I guess a number of early influences that led me to become an internet broadcaster: Jean Shepherd clearly carries part of the blame. The ham radio and short-wave listening shows of Radio Havana Cuba and HCJB also played a role. Radio shows like "Car Talk" and TV programs such as "The Furniture Guys" got me thinking about how much fun it would be to have a show about a technical or craft topic. Then the internet came along and made it possible.

We'd considered several other names for the program. I had joked with Ross that our regular conversations could be called "The Dawn Patrol." Mike and I briefly thought about that one, but I guess it didn't seem appropriate from the Alaskan point of view; our recordings sessions were pre-dawn for me, but they were night time sessions for Mike. We also considered "Radio Workbench," "Circuit Hacks," "UniJunction Junction," and "Bipolar Junction." That last one was a reference to the bipolar transistor. I guess it would also have been appropriate geographically, but it might have had some unfortunate psychological connotations. "SolderSmoke" came from the opening paragraph of my June 1998 QST article describing my Barebones Superhet project:

> *"I listened to the magic that only comes from a radio that you built yourself. In that one sentence, radio enthusiast Ian Abbott (in a message posted to an Internet mailing list) nicely described the feeling that can arise in the midst of a room full of solder smoke... and the reward that awaits those who build their own radio gear."*

Soon SolderSmoke became part of our weekend routines. At around 6 am London time on Saturday morning, I'd fire up the Echolink software and look for Mike's distinctive Alaskan (KL7…) callsign. Sometimes we'd have exchanged some e-mails outlining what we'd discuss, but usually it was completely unscripted. We'd just talk about what we'd been working on or reading about.

We used an internet mailing list called QRP-L to spread the word about the podcast, and we quickly developed a loyal core audience of listeners. The OurMedia host site reported back on the number of downloads, and soon we were getting more than 1000 per episode. At the end of each show, we asked (sometimes we begged) for e-mail feedback. It was great fun to read the messages from listeners. This became the "SolderSmoke Mailbag" segment of the show. One early correspondent, Brad Smith, WA5PSA, told us he listened while jogging through Tulsa Oklahoma. Another had listened while over the Pacific, en route to Japan. We had listeners down under, and in India. As with any of the VOIP systems, this wasn't "real radio" and it didn't have the same charm as real short-wave radio contacts, but the fact that our audience could listen to the show pretty much wherever and whenever they wanted to seemed to compensate. Lots of guys put our shows on their thumb-sized MP3 players, and listened to us on their way to work… or on their way to Japan.

Mike and I would just talk about what we were working on or thinking about, and often our discussions would lead us to other projects. For example, one day the discussion turned to crystal radios. (I think I mentioned how in his book Steve Wozniak had said that a crystal radio had given him his start in electronics.) Soon Billy and I were putting one together. It worked like a charm, pulling in a London AM broadcast station—Radio Kismat—that is aimed at the city's large Indian and Pakistani populations. The first model was built with a factory-made germanium diode, but soon we were experimenting with galena and phosphor bronze cat's whiskers, and with "Fool's Gold" (iron pyrite) that we'd bought in the nearby Science Museum. I took pictures of our extremely ugly crystal radio and posted them on our web site. I recorded the South Asian music coming out of the receiver and put the digital files on our web site.

In one episode, Mike mentioned the use of laser beams in high speed connections between computers. That got me thinking and tinkering, and soon (after some Googling) Billy and I had rigged up a laser communications system. We used a small transformer that allowed the audio output of a small broadcast band radio to vary the voltage going to a small laser-pointer. For our receiver, we used a small solar panel and the same high impedance headphones we used in the crystal radio. The transformer amplitude modulated the laser beam, and the laser beam produced an audio signal in the headphones when it hit the solar panel.

That is what SolderSmoke is really all about—sharing ideas, and talking about homebrew electronics projects. We had a lot of fun with the show. Traditions and features just seemed to pop up from nowhere. We needed a logo, so we launched a logo contest. Randy Smith, N3UMW, won the contest—his drawing became our logo. Our reading of the e-mails from listeners became a regular part of the show. I think I was influenced in this by the "DX Mailbag" feature of the ham radio program on the HCJB shortwave station. For some reason I started to bang a gong (it belongs to my kids—our sound effects resources were very limited) before reading the mail. This became an important feature of the show, and I started getting mail about the gong—some commented that I was ringing it too loudly. One guy said he missed it when, on occasion, it was unavailable (the kids would sometimes take it back) so he recorded it and sent me a digital version.

We'd been in e-mail contact with a young electronic engineer from India, Ashhar Farhan. Farhan had designed the BITX 20, a simple phone transmitter that could be built with parts that are readily available around the world. This design quickly became very popular, and hams around the globe were building them and discussing their projects on the internet. We convinced Farhan to be our first-ever guest on the show. That was one of our most popular programs. The birth of SolderSmoke roughly coincided with my decision to try to move away from the "cut and try" design techniques that had brought me so many burned fingers and so many hours of frustration. I decided that it was time for me to crack the books and teach myself some more about the design of radio frequency electronic circuitry. I was going to try to inject some method into the madness.

Amplifiers were what I really needed to learn about. I had struggled for a long time trying to understand how electrons and holes in silicon crystals are used to make transistor amplifiers. Once I got over this hump, I kind of slacked off in my quest for understanding. I suppose I subconsciously figured that if I understood how the core component (the transistor) worked, well, I didn't need to understand completely the roles of all the surrounding circuitry, all those little capacitors and coils and resistors that encircled the transistors on schematic diagrams. But it was my poor understanding of these components that was causing my fingers to get burned and my transistors to go up in smoke.

How do you determine how much gain an amplifier circuit will provide? How do you set it up so that the output signal will be a faithful reproduction of the input? When you need more than one amplifier stage, how can you ensure that the stages will work well together? All of these very fundamental issues are addressed by those components around the transistors. And that is where I now had to focus my attention.

Fortunately I already had many of the books that I needed. "Solid State Design for the Radio Amateur" had been written by Wes Hayward, W7ZOI, and Doug DeMaw, W1FB, in the late 1970s. It was at just the right level for me. During my years of "cut and try" I carried that book around a lot, jumping in and picking out bits of circuitry and understanding, but I had never really sat down with the book and tried to go through it in a systematic way. My first copy is now almost falling apart—the pages are worn and dog-eared. There are stains from high humidity in the Dominican Republic and a basement flood in the Azores. Inside the front cover there are notes from meetings with OB/GYNs and pediatricians in Northern Virginia. That book had been with me through a lot of life.

"Solid State Design for the Radio Amateur" became so popular that it is now referred to by its acronym, SSDRA. In spite of having gone through many editions, the book is now very hard to come by. If you can find them on eBay or Amazon, the prices are often above $200. Many of our discussions on SolderSmoke were about things we'd read in SSDRA. As we spoke, I came to realize how important this book had become to me... and how my only copy was deteriorating rapidly. I started to surreptitiously search the used book emporiums of the internet. Finally I found a reasonable priced backup copy. Before I sent off my order, I struggled a bit with the morality of this purchase. Is it right for one radio amateur to own TWO copies of this rare and useful book? I guess I came up with some justification for the purchase, because a much more pristine edition now sits on my bookshelf. (I later learned that I'm not alone in this—there is a guy out there who is on his THIRD copy of SSDRA— he wore out the first two.)

SSDRA's rarity is not really a problem, because it has been superseded by a new book aimed at the same audience. "Experimental Methods of Radio Frequency Design" (EMRFD) has all of the information of SSDRA, but is updated, and is a bit more focused on encouraging experimentation in the design of radio gear.

Author Hayward is seen here during a mountain trip on which he took the 40-meter Ultra Portable Transceiver described in this chapter. The battery pack is in his jacket pocket.

SolderSmoke brought us into contact with the one of the authors of SSDRA and EMRFD, Wes Hayward, W7ZOI. Wes is a legendary figure among melters of solder, and I must say that Mike and I were both more than a little intimidated when e-mails from him started coming in. We'd both been reading his books and articles for years. Among the technical writers, Wes has a very distinctive persona. He is from the West, and over the years he has combined his love for circuitry and radio with his passion for the great outdoors. There is one picture of Wes in SSDRA. It is in the "Field Operation/Portable Gear" chapter. He is standing in the woods, upright and confident in a ski cap and winter jacket. The caption reads, "Author Hayward is seen here during a mountain trip on which he took the 40 meter Ultra-Portable Transceiver described in this chapter. The battery pack is in his jacket pocket." On the page opposite is a picture of a neat and very unusual ham radio station. "Photograph of the W7ZOI home station. All of the amateur equipment and test gear is home made. The operating position serves double duty by also being a workbench." Battery pack in the jacket pocket... All gear homemade... Even Serge from Kiev University would have felt like an appliance operator in the presence of Wes Hayward.

There were other inspiring pictures in SSDRA, but unlike the two mentioned above these others were inspiring because of their ugliness. This requires some explanation: When I was a kid trying to get started in electronics, the things that I built never looked even remotely like the neat projects that appeared in the ARRL handbooks or in the pages of QST. Inventions presented in those publications always had their components neatly arranged. All wires were bent at crisp 90 degree angles. All solder joints looked like they had been made by NASA specialists in some sort of clean room. No, my products did not look like that. My products were ugly. But so were the projects in SSDRA and EMRFD. The "General-purpose broadband push-pull amplifier" pictured on page 63 of SSDRA looks like it could have come straight out of the 1972 Congers, N.Y. laboratory of WN2QHL. The two transistors are, well, askew. No neat right angles here. Components float in the

air, suspended by their wires and attached to other floating components by blobs of solder. Here was an electronics book that I could identify with. In July 2007 Wes and I happened to be in San Diego on the same day. We met up for breakfast and I got him to sign my old copy of SSDRA.

I began my effort to understand amplifier design shortly after our arrival in London. My tools in this effort were SSDRA, EMRFD, a few other books, a calculator, and a spiral notebook. My mentors were on the internet. My classroom was my London radio shack, but I took my learning tools with me on all of our family trips: U.S. military bases in East Anglia, UK; Cape May, N.J.; Santo Domingo, Dominican Republic; the beaches of South West France—in all these places I struggled to understand the design of solid state amplifiers.

UNDERSTANDING: AMPLIFIER LOADS

On my old DX-40 transmitter there was on the front panel a very prominent and obviously important control knob labeled "LOAD." Together with another knob called "TUNE" it was used to make critical adjustments to the transmitter's final amplifier. Adjust these two controls properly and your signal would go up to the antenna out to the universe. Adjust them incorrectly and you could end up transmitting harmonics and risk a visit from FCC-men in dark suits. Or—worse yet—you could "blow your final" and knock yourself off the air.

I knew that these two knobs controlled the settings of capacitors in the resonant circuit of the 6146 final amplifier tube. And I vaguely understood that one of the things they did was to determine the impedance—"the load" to put it in mechanical terms --- that would be faced by the final amplifier.

Think of a bicycle. Adjusting the gears varies the "load" faced by the rider's legs. You want to get it just right. If the load is too light you find yourself flailing away at the pedals, pedals that are too easy to push. If the gears are adjusted for too heavy a load, you can't even turn them around for one rotation. Ultimately, my antenna system was my load. The "Load" and "Tune" knobs were the equivalent of the gear controls on my bike.

That's the general idea, but I found it difficult to translate this concept into an understanding of solid state amplifiers. My problem in this is a good example of how the presentation of a mathematical formula (without explanation) will often be insufficient to bring real understanding to many readers.

What should the load be for an amplifier? 50 ohms? 5000 ohms? 2 ohms? How can you know? Why is this important for amplifier operation? I knew from bitter and painful experience that it was important, because I strongly suspected that many of the transistors that I'd burned up were fried because of inappropriate loading. But when I went to some of the books to try to get a sense of how to do this right, all I got was this:

$$R_{load} = (Vc - Ve)^2 / 2P_{out}$$

In words, this equation says that to determine the load for your amplifier, take the voltage across the collector and the emitter, square it, and divide it by twice the power output that you desire. I guess some people would be inclined at this point to say "Voila!" or "Of course!" I, however, found myself scratching my head and muttering, "Huh?"

I had no doubt that the formula was correct, but I just didn't understand the logic and science behind it. For me, it was just another black box, in this case a mathematical black box, a math appliance. I could feel Stan, and Bollis, and Serge, and C. Bettencourt Faria all looking over my shoulder, thinking, "You don't understand this, you APPLIANCE OPERATOR!"

So I needed to understand *why* $R_{load} = (Vc - Ve)^2 / 2P_{out}$

Let's review some of the Ohm's law equations and substitutions that were explored by Richard Feynman in his boyhood lab:

E=IR or I=E/R or R= E/I

P=IE or I=P/E or E=P/I

Through substitution:

P=IE becomes P=(E/R)E or $P = E^2 / R$

P=IE becomes P=I(IR) or $P = I^2 R$

If $P = E^2 / R$, then $R = E^2 / P$

We are now getting close to the equation that had me scratching my head.

Earlier we said that an amplifier could be thought of as a variable resistor controlled by the input signal. Obviously changes in the resistance control the amount of current. But you also need to remember that the voltage <u>across</u> that resistor will vary as the resistance changes. At one extreme, with the variable resistor at maximum value, it will look like an open circuit—if you put a voltmeter across it, you would measure the full value of your voltage source, your battery. At the opposite extreme, the resistor will look like a short circuit. It will appear to be a piece of wire. If you take a voltmeter and put both probes on the same piece of wire, well, you will not measure any voltage between them. So, as the value of the variable resistor changes, the voltage across it varies from the full battery voltage down to zero. It is easy to see how the *current* through the variable resistor changes, but you have to keep in mind that the *voltage* across it is varying too. That kind of variation in both current and voltage is one of the keys to understanding that mysterious load formula.

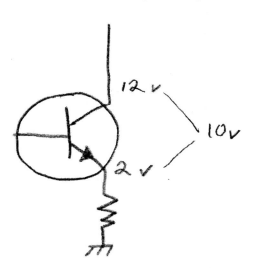

Let's assume that we are designing a *linear* amplifier, one in which the output is a faithful, but hopefully more powerful reproduction of the input. This will be a Class A amplifier.

Vc-Ve means the voltage on the collector, minus the voltage on the emitter. This is simply the voltage across the transistor when the transistor is off, the equivalent of the battery voltage in our example above.

We're trying to determine what load value to use for a desired output power with a given voltage value. Let's look at that voltage value. Assume that we have 12 volts on the collector and 2 volts on the emitter. So there are 10 volts across them. What is the maximum peak value of *signal* voltage that we could get out of this amplifier?

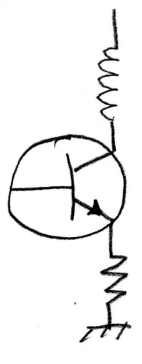

The amplifiers we use almost always get their DC voltage through either an RF choke or an RF transformer. Because energy is stored and released in these coils, the *signal* voltage at the collector can actually rise above the DC voltage. But there is a limit to how far we can go with this. Remember, we want linearity, we want the output wave form to look just like the input wave form, only stronger. So that voltage on the collector can swing up beyond 12 volts, but when the input signal (think about turning the control on the variable resistor) pulls it down, it can only go down so far. In this case it can only drop by 10 volts, because beyond this point there will be no voltage difference between the collector and the emitter, current through the device will be at maximum, and the output will no longer faithfully follow the input signal. So, the most we can do is set up the circuit so that the voltage on the collector will be starting at 12 volts, rising to 22 volts (when the input signal is pushing the transistor to maximum resistance, current is approaching zero, and the coil or transformer is providing additional voltage) and then dropping to 2 volts (when the input signal is pushing the transistor to minimum resistance and maximum current is flowing through the transistor).

We're now looking at an RF signal superimposed on a DC voltage. We get rid of that DC by running the signal through either a capacitor or a transformer. We end up with an RF signal varying from zero, going up to 10 volts positive and down to 10 volts negative. We say the peak signal voltage is 10 volts. This is the value that we plug into the formula that we are trying to understand. Note that it is the same value as (Vc-Ve).

We now know how much voltage the output signal can have. And we know that Power = Voltage X Current. With a given voltage, Ohm's law tells us that current is determined by resistance. So if we want a specific output power level, and we know what the maximum voltage value will be, we can use Ohm's law to pick a resistance value that will yield the desired amount of current.

We've explained most of the formula that had me scratching my head, but not all of it. Why is there a 2 in the denominator? Where did that 2 come from?

When we talk about power in an AC circuit, we are talking about average power. Voltage and current are varying throughout the cycle. We don't just pick the peak values and do the math using them. We want average power. To make a long story short, putting a 2 in the denominator simply turns our peak voltage value (10 volts) to a value that in our equation yields AVERAGE power.

So what this formula $R_{load} = (Vc - Ve)^2 / 2P_{out}$ is saying is this:

The voltage from the collector to the emitter sets a limit on the maximum signal voltage. Power is determined by both voltage and current. Rload sets the current.

If you have a maximum signal voltage of 10 volts (Vc-Ve) and you want 2 watts delivered to the load, you should put a load of 25 ohms on the final transistor. You can check this out by going back to the simple $P = E^2 / R$ formula.

$(10 volts)^2$=100volts,   2*25 ohms=50,   100/50=2

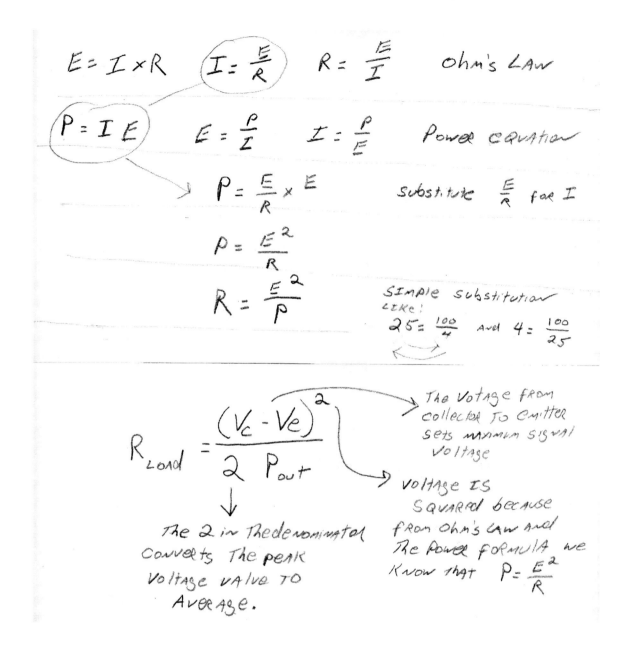

Another way of doing this would be to convert the peak value to an RMS value. RMS stands for Root Mean Square and is a way of calculating the DC equivalent of an AC signal. To get RMS you multiply the peak value by .707. If you use an RMS value for the voltage, there is no need to put a 2 in the denominator:

Using an RMS value for the voltage:

10 volts x .707= 7.07 volts, 7.07^2 = 49.9849, 49.9849/25 ohms =1.999 watts

Of course, this assumes that you are going to be driving the transistor with an input signal sufficient to cause the voltage in the output circuit to swing to its maximum value. With a given value of load, if you drive the stage with a signal that does not cause the output signal voltage to reach its full value, you will have lower levels of output.

****

Prompted largely by the conversations on SolderSmoke, around this time I started thinking that computer assisted design tools might help me understand and design circuits. Circuit simulator programs had been around for many years, but the old 286 and 486 computers that I had been using left most of these programs out of my reach. Articles in QST praised these programs as marking the end of the "cut and try" approach to homebrewing, but with my old clunky computers I was left cutting and trying. Then in 2006, it finally became very clear that I needed a new computer. I'd written a book (sadly, it is not about radio) and had found a publisher, and I knew that I'd need a machine that would support all the back-and-forth with editors that would precede publication. So I got a new computer. This opened the door to SPICE.

SPICE stands for "Simulation Program with Integrated Circuit Emphasis." It was developed during the 1960s on the campus of the University of California at Berkeley. An earlier version of it was known as CANCER for "Computer Analysis of Non-Linear Circuits, Excluding Radiation." (The "Excluding Radiation" part was apparently a 1960's era protest directed against the U.S. Defense Department—thinking of nuclear war, their contracts had required other circuit simulators to be able to evaluate the radiation hardness of the circuits being simulated.)

As the name indicates, SPICE was intended for the evaluation of integrated circuits. The manufacturers of these expensive and extremely complicated chips could not afford a lot of cut and try. They needed a tool that would help them determine with some certainty if a proposed design for an IC would actually work. SPICE and programs like it gave them that predictive ability. In essence, SPICE allowed them to "build" a new chip in a simulator. All of the electrical and physical characteristics of the various elements going into the chip would be included in the program. SPICE would then—in a virtual way—combine all these components in the way intended by the designer and predict how the chip would actually function.

Fortunately for radio amateurs, this kind of software could also be used to evaluate the kind of simple, discrete component, analog circuits that we worked on as hobbyists. Around the time I got interested, a very user-friendly version called LTSpice (Linear Technology Spice) became available.

The student version is free and downloadable. The user is presented with a blank schematic design page. It is reminiscent of a blank piece of circuit board. Virtual "parts bins" line the top of the screen. You click and drag the desired parts onto the schematic page and begin to assign them values and wire them together.

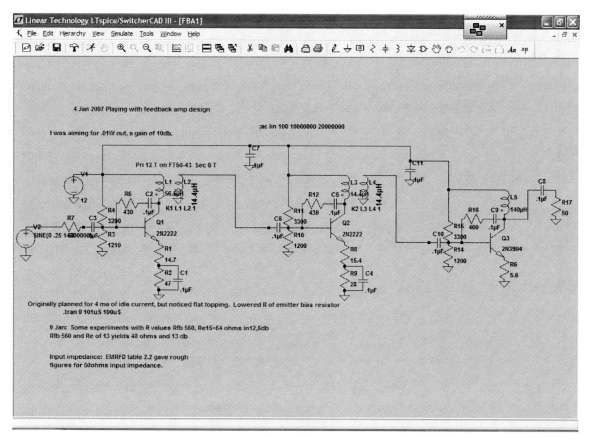

Once the circuit is completely drawn, the real magic begins. There is a little "running man" in the upper left—that's the icon used to run the simulation. Click on him and the computer takes your schematic and makes it run in the virtual world.

How do you know how it is running? Well, LTSpice comes with a complete set of test instruments. Select the oscilloscope probe, for example, and you can see what the voltages look like at any point in the circuit. With this tool you can easily see if that amplifier you just designed is distorting the signal. You can measure the power gain. You can measure the input impedances. You can look at the spectrum at the output to see if there are any signals there that should not be there.

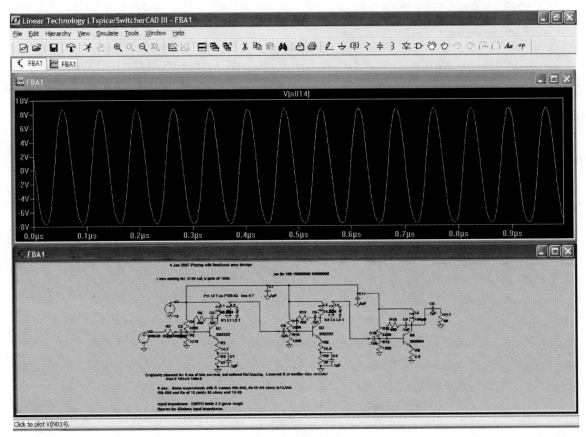

Even for relatively inexperienced homebrewers, LTSpice is a tremendously useful tool. When you are doing the kind of hacking and circuit combination that many of us do, LTSpice provides the opportunity for a kind of electronic reality check. Very often, when you build a new circuit on your workbench and it does not work as planned, the reasons for the failure are very unclear. It could be that you made some error in constructing it. Or it may be that the "design" you came up with defies the laws of physics and could never possibly work. LTSpice lets you check on this. You take your design and "build" it in LTSpice. Hit the running man icon and see what the computer predicts. If it shows the kind of output that you expected, then you know that your design is at least theoretically feasible.

LTSpice is also very useful when you are trying to understand the principles behind electronic circuits. When I was struggling to understand the operation of transistor amplifiers, for example, I could do a virtual build of the circuit that I was studying. I could then easily play with the values of the components around the transistor and instantly see the effects of the changes on the gain of the stage, the current through the transistor, the linearity of the stage, etc. Somebody on one of the homebrewer internet mailing lists observed that, together with books like "Experimental Methods in RF Design," LTSpice can be seen as a veritable "BSEE in a can"—you can learn a lot with this program.

Mike and I were enthusiastic users of LTSpice. Soon we figured out how to make videos of our simulation sessions. We called these videos "Spice-casts" and posted them on our web pages and on YouTube. I'd been making videos for a while—each time my family and I took a trip, I'd record a short video message for the SolderSmoke gang. I urged Mike to do more video, but he seemed camera shy and never showed his face on his Spice-casts.

My effort to better understand the circuitry took place at the same time that the SolderSmoke podcasts were starting to gain an audience, and feedback from our listeners became an important resource in my effort to educate myself—I'd talk with Mike about a particular bit of theory that I was struggling with, and soon I'd be getting helpful e-mails from SolderSmoke listeners who wanted to shed light on the problem.

It soon became apparent that one of the reasons I had such a tough time getting homebrew transmitters to work properly was that my test gear was very limited. Wes Hayward pointed out that the ability to accurately measure power levels was very important—you really need to know how much power you are putting into your amplifier stages. He noted that the very simple RF probe that I was using (really just a germanium diode, a capacitor and a resistor) was not capable of accurately measuring low levels of radio frequency energy. He recommended the construction of a power meter that he and Bob Larkin, W7PUA, had designed.

I knew I needed one of these meters, but I had somehow developed an aversion to the construction of test gear. "I'll build the radios," I said, "but I'll leave the test gear to the commercial manufacturers." I guess I feared entering some sort of homebrew vortex in which I'd find myself endlessly building test gear that was needed to build and test other bits of test gear. Don't laugh. It could happen. Test gear construction can get out of hand. When I was building the Doug DeMaw superhet receiver, DeMaw recommended the use of a bit of test gear that would help select crystals for the filter. As I recall, that bit of test gear was more complicated than the receiver itself.

But for some reason, I seemed destined to build this power meter. Mike knew I needed one, and several times offered to build one for me. I didn't want to burden him with yet another technical project, so I kept saying that I'd build my own. The meter is designed around an Analog Devices AD8307 chip. It is a very sensitive logarithmic amplifier. After hearing Mike and me go back and forth several times on my need for the meter, one of our listeners in Norway, Thomas (LA3PNA) very generously put the needed chips into an envelope and sent them to me in London. In doing this, Thomas was carrying forward the long ham radio tradition of sharing parts with fellow amateurs. He was acting on the same feelings of ham radio brotherhood that caused those American hams to send C. Bettencourt Faria the crystals that he needed to build his rig in Angola in 1958, and that caused Tom, W1HET, to send me "care packages" of parts when I was in the Dominican Republic. This attitude of "my junk box is your junk box" is one of the really wonderful things about this hobby. The internet seems to be strengthening and globalizing this fine tradition—the casual mention of a need for parts on the QRP-L mailing list, for example, will often result in unsolicited parts packages from the other side of the planet showing up in the mailbox.

Chips in hand, I dug up the QST article by Wes Hayward and Bob Larkin (it was on the CD that came with EMRFD) and started rummaging through my various junk boxes and soon found further evidence that the radio gods wanted me to build this meter. There in bottom of one box of radio treasures was a pristine aluminum box of exactly the right dimensions. The only thing I needed now was the actual meter that would display the logarithmic findings of the AD8307. If any of you harbor any doubts about the influence of the radio deities that I have been mentioning, well, how do you explain the presence in my junk box of EXACTLY the same kind of RadioShack meter movement called for by Hayward and Larkin? It was just sitting there, one of only three or four old meters in the junk collection, still happily protected by its Radio Shack packaging. Like I said, there are some projects that you are just destined to complete.

One of our other guests on SolderSmoke, Jim Burrell, N5SPE, once commented on the feelings of contentment and happiness that come from just being in the workshop with a cup of coffee on a weekend morning. My son Billy and I were in that happy zone several times during a long weekend in May 2007 in London. As he worked on one of his projects, I put the W7ZOI meter together. I managed to drill the needed holes in the box without mangling it too badly (metal work is still not my strong suit), and somehow the chips survived my less-than-scrupulous attention to the static electricity threat (you're supposed to keep yourself grounded while handling the chips, but I like to live dangerously). I was just about to finish, when I paused and called Billy to the bench. I had him solder in the last component.

Billy had helped me wind a toroidal coil when he was just two years old. I don't push my kids into ham radio, but I do like it when they help me with my projects; this kind of assistance definitely increases my feeling of personal connection to the end product. During our last weeks in London, Maria watched me work on a power supply that I was refurbishing. I'd picked it up at the Kempton Park hamfest, and had rebuilt the circuitry using a schematic from a book by Tony Fishpool, G4WIF, and Graham Firth, G3MFJ. I explained to Maria that I didn't like the color of the cabinet, and that I was going to paint it. Maria asked to help. So out to the garage we went with paint can in hand, and Maria painted that cabinet.

Many years ago Tracy Kidder wrote a book with the wonderful title "The Soul of a New Machine." Parts donated by friends, ideas and inspiration from other friends, serendipitous junk box parts, abstruse circuit concepts that suddenly become clear amidst a cloud of solder smoke, components soldered in by your son, a cabinet painted by your daughter… these are the kinds of things that add soul to a new machine, the kinds of things that make a shack a place filled with happiness and contentment.

But of course, even the happiness and contentment of a Saturday morning in the ham shack can be disturbed by the sad parts of life in this world. My friend Mike did not live to hear about the completion of the power meter that he'd told me I needed. One sad morning in January 2007, I'd climbed up to the attic shack with coffee cup in hand and checked my e-mail. There was a message from a SolderSmoke listener in Hawaii. He forwarded a newspaper article about a car accident that had claimed the life of a 51 year-old tourist from Alaska. It was Mike. He'd gone to the islands with his wife and son. They survived with only minor injuries, but Mike was killed.

Mike's death came as an enormous shock. It was my sad duty to pass the news of his death to the worldwide fraternity of solder melters. I posted a message on the QRP-L mailing list, and put a note about the accident on the SolderSmoke web page that Mike himself had first set up.

People were really stunned. I think hams around the world had grown used to getting nothing but happy, friendly news from SolderSmoke and the associated web pages. Mike and I had assiduously avoided any controversial or unpleasant topics—there was no discussion of politics or war. We even avoided ham radio topics that were likely to cause discomfort (the debate about the future of Morse code, for example). And while our conversations had allowed thousands of people around the world to get to know us, the program—like almost all ham radio conversations—gave a somewhat filtered version of our lives. They were ham shack conversations, and the sad realities of life (and death) are usually kept out of the ham shack, kept out of the contacts. Suddenly when hams around the world turned to the web sites that usually brought them happiness and fellowship and light-hearted tech talk, they found instead very bad news, very sad news. Mike was gone.

And of course, the wave of shock and sadness that swept through our little (but global) community was the direct result of the kind of guy Mike was. He was a very nice person, and this came through in every one of the forty-six shows that we did together. Even people who were not very interested in the technical subjects that we were discussing tuned in to SolderSmoke just to hear Mike's calm and pleasant voice talking about his projects.

Of course there was a real possibility that Mike's passing would be the end of SolderSmoke. But in the many condolence e-mails that I received, I noticed that a good portion of them contained pleas for SolderSmoke to continue. Many people said that the show had become a part of their routine—some listened to it while jogging, or while on long commutes. Others took it with them on business trips or just liked to listen while working in their shacks. They were telling me that SolderSmoke would be missed.

I gave it a lot of thought, and considered what Mike would have wanted. Just before he had left on that fatal trip to Hawaii, Mike had told me that I should go ahead and make the next show without him—he didn't want our listeners to miss an episode just because he was on vacation. I told him the show would wait for his return to Alaska… He never made it back, but his desire to have the show go on was quite clear. Even without this parting comment I knew that he would have wanted SolderSmoke to go on. It was his creation as much as it was mine. He knew that people liked it. I could almost hear him telling me to "make another one." So, after a couple of months, I recorded the first solo edition of SolderSmoke.

At around the same time, the familiar storm clouds were gathering. I started getting my "Foreign Service nightmare," that dream about opening a door that had been ignored. It was time to leave London. We were going through the nerve-racking process of negotiation with the State Department about our next home. For a while it looked like we would be going back to Santo Domingo. Then it looked like it would be Madrid. Then, one night right after a dinner in which we'd all fretted about where we would go, my BlackBerry started to buzz. There was an important message: "Come to Rome."

CHAPTER 7
ROME
SECRET RADIO IN THE ETERNAL CITY

"To invent, you need a good imagination... and a pile of junk."
—Thomas Edison

*"...he could not get rid of a chronic disease, the **cacoethes gadgetendi**, the itch to tinker."*
—Author Richard Preston describing astronomer/gadgeteer James E. Gunn in "First Light – The Search for the Edge of the Universe."

I was a bit nervous. "Right in front of the main entrance of the U.S. Embassy" had seemed like a good place for us to meet, but as I waited I had my doubts. It is not really a friendly spot: The policeman with the sub-machine gun usually gives you a "Buon giorno," especially if you work in the building... but he is still toting that sub-machine gun. And I wasn't being very discrete either: as I stood there with that black computer case in my hand, I could almost hear people wondering, "What is he doing out there? Why is he carrying that bag at lunchtime? What's in that bag?" And of course, I was nervous precisely because of what I had in that bag.

The JNI guy was a couple of minutes late, and I began to suspect that we might have had a misunderstanding about the meeting point. The Embassy is a very big place and my Italian was still quite shaky, so I thought he might have misunderstood, and might be waiting at another entrance. I started walking around the block looking for him, getting the hairy eyeball from each policeman and security guy that I approached. I could almost hear the questions in their heads: "WHAT DOES HE HAVE IN THAT BLACK BAG?"

Finally, my cell phone rang and I learned that the JNI man was up the street a bit. He was easy to spot—like me he had his cell phone to his ear and he was looking around for someone.

Quickly we moved away from the sub-machine guns and security cameras and went to a small café across the street. As the JNI guy ordered the coffee I got us a table and prepared to unveil the device.

First I showed him a picture of it. "Here it is," I said.

He was shocked: "Mio Dio! That's it?"

Then I put the picture away, and I prepared to really surprise him.

"I have it with me."

Really? Is it... in the bag?"

I smiled and carefully opened the flap on the computer case. The JNI man's eyes widened as I pulled out the book-sized, copper-colored box with multiple switches and knobs.

"There it is Fabio! That's the little Double Sideband rig that I was using when we met on 20 meters!" Fabio Galeffi, IK0JNI, was truly astonished by the small size of the homebrew rig that had brought us together. I guess I bumped up against the limits of coffee bar decorum when I pulled out a small screwdriver and opened the case to show him the circuitry.

Over coffee we filled out the QSL cards that confirmed our contact of the previous evening. And we talked about the international fraternity of radio amateurs.

It was very appropriate that th... st meeting with Fabio had some undercurrents of clandestinity, because all of my ... ...perations in Rome had an air of secrecy about them. Not only was t... ...o keep my life as a radio fiend secret from my diplomatic colle... ...yself in a small apartment surrounded by somewhat finicky neigh... ...m know that they now had among them someone with a very bad ca...

The... ...onceal. Our apartment in Rome is a bit small, so some downsi... ...ry. Before we left London I'd sent quite a bit of stuff into storage... ...nd hopefully at that point I'll have a shack large enough to accomm... ...hack in a small room off the kitchen that had originally served a... ...ed, but I think this kind of accentuates the "radio shack" atmosphe... ...y the intrepid radio pioneers of yesteryear (in backyards and on sh... ...d I liked having the shack close to the kitchen—not only was the c... ...kitchen was three levels below me, so far away that I carried TV... ...s each morning), but—most important—this kept the family clos... ...o in and see what I was up to. I could call over to the kitchen and... ...onders of my latest technical marvel (she is very good about feignin... ...nd wonder). When necessary, I'd just close the shack door and vis... ...lly ignorant about the presence of THE KNACK in their midst.

My first ant... ...tely invisible. I used 30 gauge-enameled wire—this stuff is not mu... ...ged to get about 40 feet of it stretched between our living room wi... ...from a patch of tough, urban Roman grass that is used mostly for... ...the dogs would actually break my antenna). From the window, I ta... ...g the molding on the floor. My antenna tuner (T network) was ma... ...rs and a coil wound with bell wire on one of the pill bottles that I h... ...get receiver project.

As I was worki... ...was also trying to get the Italian authorities to issue me a reciproc... ...ave a very formal diplomatic agreement in this area, but I quickly i... ...cense was quite difficult. Even with diplomatic status and help fro... ...eople, what had been a simple and painless process in other cou... ...ratic nightmare in Italy. But with the diligent help of Piero Ippolito... ...h.

It was early Februa... ...ne together into a fully authorized radio and antenna system. At th... ...usty Heathkit HW-8 QRP transceiver and, on a rainy Roman winter afterno... ...that the hodgepodge of wire hanging invisibly from my window would actually tune up on almost all of the amateur HF bands. My first QSO was with G4BJM on 20 meter CW. I was back on the air.

Talking to my old home in the UK was great, but I was really looking for contacts closer in. I'd regretted not making more friends among the British hams, and had vowed to do better in Italy. So when I heard an extremely strong SSB signal on 20 meters one afternoon, I paid close attention to the callsign, and immediately Googled it. What luck! Google revealed that I0ZY was Gianfranco Scasciafratti, a well known DX champion. And he was located within a mile of our apartment. The QRZ.COM web site provided an e-mail address. Soon we were in touch, and Gianfranco invited me to a Saturday afternoon meeting of the Rome chapter of ARI, the Italian amateur radio association.

We met in front of the Education Ministry (you really can't miss it). Gianfranco introduced me to Giorgio Castelnuovo, I0YR, his neighbor and a fellow DX fiend. As we drove to the clubhouse we

> *And the peace of God, which passeth all understanding, shall keep your hearts and minds through Christ Jesus.*
>
> Philippians 4.7 (KJV)
>
> AMERICAN BIBLE SOCIETY
> AmericanBible.org

talked about ham radio. Giorgio had been in the hobby for a long time—he told me his first license was issued by U.S. military occupation forces during World War II.

The club house was very similar to that of the Crystal Radio Club, and the Dominican Radio Club, and the Azorean Radio Club: Piles of old magazines. Bulletin boards with pictures of long-ago picnics and Field Days. Mailboxes for incoming QSL cards. Some old gear. It was great. I felt right at home.

Gianfranco said we couldn't stay long, because Giorgio desperately wanted to work on his antenna. It was a very big quad (big square loops of wire in the sky—picture "the Ray-Gun" on steroids), and it had been knocked down by a windstorm. Feeling the call of the radio amateur's "prime directive" ("Help Other Hams!"), as we drove home I volunteered to assist. Soon we were up on Giorgio's roof, wrestling with a really huge fold-over tower. We joked that my 30 gauge wire antennas were completely at the other end of the spectrum.

After a few maneuvers that caused me to think about the status of my life insurance policy, we had Giorgio's tower in the vertical position. (Giorgio had done most of the heavy lifting—I was really surprised a few weeks later to learn his age. He's 81! He tells me that his interest in ham radio keeps him going strong.) It had been a wonderful afternoon. I headed home feeling like I'd become a part of the ham radio world of Rome. It was really great—I'd finally met up with some local hams. I'd been to a club meeting. I had an Italian license, I had an antenna up, and I was ON THE AIR…

But then my HW-8 started to act up. It's always this way: Just as you seem to get your technical act together, just as you get one part of your station working right, another critical part will go bad on you. It's almost enough to make you take up stamp collecting.

When I was a kid, these kinds of problems really drove me nuts. Now they don't bother me so much—I guess it is a combination of improved technical abilities, and the increased patience that comes with getting older. Also involved is a savoring of the challenge of troubleshooting—the thrill of the hunt.

This particular HW-8 problem was a difficult one—it was intermittent. Sometimes, I'd be happily pounding out Morse code on my straight key, when I'd look up and see that the SWR/Power meter that normally kicked up whenever the key went down, well, didn't. This was obviously not good. Out came the manual and the schematic diagram for the old Heathkit. I was kind of hoping that it wouldn't be something simple—sometimes problems like this are resolved by a squirt of contact cleaner on a dirty relay connection. That's easy. Too easy. In this case it soon became clear that the HW-8 was throwing up a greater challenge.

A review of the symptoms (transceiver switches properly from receive to transmit, but power out intermittently drops to zero) caused me to focus on the circuitry around the driver amplifier and the final amplifier. I checked the voltages on the transistors, and found that when my output dropped to zero, there was no voltage on the collector of the driver transistor. That'll do it! Next I learned that a PNP switching transistor—Q11—was used to put voltage on the driver when the transceiver is in transmit mode, and take it off when in receive. Q11 was the prime suspect.

Now it was time to test my theory. The soldering iron was heated up and soon Q11 was plucked from the printed circuit board. A bit of wire was bridged across the gap that should have been opened and closed by Q11's switching action, and the circuit worked perfectly. Q11 was clearly the problem.

Now I had to look for a replacement part. I turned to my vast and somewhat organized junk box. (I keep most of my parts in plastic boxes that originally contained baby wipes.) At first I thought that I might be stuck without a suitable part, but there in the bottom of one of the boxes was a PNP switching transistor just waiting to be used. In it went, and the HW-8 was immediately working as the wizards of Benton Harbor had intended.

There aren't many things in this hobby that provide as much satisfaction as building a completely homebrew rig, but I often think that the joy of a good repair job comes close. Taking something that doesn't work and putting it back in operation through the application of technical knowledge and workbench skill is a very satisfying thing.

Paul Huff, N8XMS, captured this feeling of satisfaction when he described his trouble (as a Novice) with an ancient Heathkit AT-1:

> *One day the AT-1 stopped working. Being more bold than intelligent, I unplugged the rig, opened it up, and started poking around inside. I had no schematic and almost no idea of how the thing was supposed to work, but I eventually spotted an RF choke that looked like it was a little bit charred. Without knowing anything about component values I went to my junkbox to look for a replacement. My junkbox was literally a box filled with old telephone parts, TV chassis, and broken transistor radios. I found a choke that looked about the same - at least it had the same number of honeycombed coils of wire along its length. So I got out my soldering gun and quickly replaced the old part. After closing up the rig I fired it up and bingo, I was back on the air! That blind luck gave me such a feeling of accomplishment, and I knew then that I was a "real ham!"*

That's why my old friend at the Virginia hamfests used to say "I pay extra for that" when vendors admitted that a rig wasn't working. It turns out he wasn't the only one who felt this way. Writing in EDN magazine, Jim Williams, an engineer with Linear Technology Corporation wrote:

> *The fact that the instrument is broken provides a unique opportunity; a broken instrument (or whatever is at hand) is a capsulized mystery, a puzzle with a definite and very singular "right" answer. As a result, you are forced to measure your performance against an absolute, nonnegotiable standard. When you're finished, the thing either works or doesn't work... And, finally, fixing is simply a lot of fun. I'm probably the only person at an electronic flea market who pays more for the busted stuff than for the equipment that works! Oh boy, it's broken! Life doesn't get any better than this.*

My revived HW-8 got me back on the air, but it was the Double Sideband transceiver that I showed to JNI in the coffee shop that served as my magic carpet in Italy. It was that rig that put me in contact with unmet friends.

Soon after the HW-8 problems were sorted out, I had decided it was time to try to make some voice contacts. I had hooked up my little homebrew 20 meter double sideband rig. Soon I'd heard Fabio talking to another Italian amateur, and at the end of their conversation, I gave him a call. This became my first HF voice contact from Rome. I explained to Fabio that I was testing antennas and homebrew radio gear. He immediately volunteered to help out with on-the air tests. We set up a schedule for morning and early evening contacts.

During one of our evening contacts, Giorgio came on frequency. Then we were joined by Giorgio's neighbor, Walter Filippi, IK0ZMH. A few days later Gianfranco was with us. Soon it was decided that 8:30 p.m. was the optimum time for us to meet—this fit in well with Italian mealtimes, and with the bed times for my kids.

These "otto e mezzo" contacts quickly became an important part of my daily schedule—after dinner, I'd step into the shack to say hello to the gang. We'd compare notes on the day, and talk about our gear. Once in a while Maria or Billy would say a few words. Through these 8:30 contacts we strengthened our friendships. For me, these contacts were made even more pleasing because I was using homebrew gear.

But it quickly became apparent that my homebrew gear needed some work. I'd built that DSB transceiver six years earlier. I'd thrown it together with reckless disregard for some pretty fundamental radio design principles. Like impedance matching. Like gain distribution. Like harmonic suppression. At the time, these things seemed like troublesome details—fancy frills that might delay my triumphant homebrew arrival on the 20 meter phone band. I just took bits of circuitry and slapped them together. Sometimes I actually cut out (with scissors) portions of schematic diagrams, and then scotch taped them to other bits of schematics. It was like electronic kindergarten.

But now I was older and wiser. I'd read "Solid State Design for the Radio Amateur" (SSDRA) and "Experimental Methods in RF Design" (EMRFD) and I'd been mentored via the internet by 'ZOI himself. I'd downloaded LT Spice. I'd bought more test gear. I'd reformed. I'd changed. I'd moved forward. But like the ancient ruins that appear throughout the Eternal City, the homebrew rigs that surrounded me in my Roman shack stood as reminders of a brutal and unenlightened past.

The signals these rigs put out were often weak and embarrassingly ugly. The 20 meter DSB rig was no exception. The weak part didn't really bother me. I am, after all, a QRP man. But the ugly part gnawed at me. As a kid, I guess I'd really internalized the notion that a true radio amateur should take pride in the quality of his signal. It wasn't just the fear of swift justice from agents of the FCC (real as that threat seemed to be). It was deeper than that. At the very least, a bad signal was a reflection of poor technical ability—if your signal had defects, each time you pressed that telegraph key or push-to-talk button, you would be providing the world with evidence of your technical weakness. It was almost a question of character: key clicks, distortion, harmonics, RF feedback, AC hum—these were the shameful calling cards of electromagnetic disrepute.

When I'd thrown these rigs together, I was happy to just get them on the air. The lack of signal purity did bother me, but I put it to the back of my mind. I didn't fret too much about the fine points. But by the time I got to Rome, I wanted more. I wanted to clean up my act. I wanted clean signals. I wanted output on just one frequency at a time. I wanted my fellow amateurs to marvel not only at the astonishing fact that I was speaking to them with homebrew gear, but I also wanted them to be gobsmacked by the purity and clarity of my RF output.

So, in an effort to make listening a bit easier on my new Roman friends, and to make me feel better about myself as a homebrewer, I started working on my old homebrew relics, giving them electronic design overhauls.

First up for overhaul was the 20 meter DSB rig, and the first step was drawing a schematic of what I'd built. This was like reverse engineering. Or ham radio archeology. That rig was a bit like Rome itself: There were many layers of development, with new stuff being placed atop the old. And I had to dig down to investigate to find out what had been done in the past.

I now had a good L-C meter (from Almost All Digital Electronics) so now I could pull out coils and capacitors and measure their values. The W7ZOI/W7PUA power meter gave me an accurate way to measure the power outputs of amplifier stages. And of course the HAMEG scope from the Kempton Park Rally allowed me to peer into the circuit and see the waveforms. After each bit of the schematic was done, I'd often draw the circuit in LTSpice. This would give me an easy way to check on performance and on the possible usefulness of any ideas for improvement. All the new gear and software helped, but I think the most important tool in this radio reform effort was the enhanced understanding of the circuitry gained through EMRFD, SSDRA, and the online mentoring of Wes and other wizards. Like Feynman said, you really fix radios by thinking.

I'd thrown together that 20 meter DSB rig in the Azores in those heady days following my success with the 17 meter DSB rig. In fact, it had first been built for 17 meters; I'd changed it to 20 meters when the sunspots started to disappear. The circuit architecture was much the same as that of the 17 meter rig: a variable crystal oscillator (VXO) as the only common stage, with direct conversion receive circuitry on one side of the PC board, and a simple balanced modulator to RF amp transmitter on the other side. The original RF amp circuit was based on a Doug DeMaw article in CQ magazine. (His had been carefully designed for 40 meters, but with my typical reckless abandon I'd put it on 17 and then on 20.) I got the idea for the receiver from a SPRAT article penned in Australia; Bill Currie, VK3AWC, had used three NE602 mixer chips and one audio amp to build a DC receiver. It had an NE602 as front end RF amplifier, another as the actual mixer (product detector) and a third as an audio amp. An LM386 chip amplified the audio to headphone level.

Right from the start, the receiver worked great, but the transmitter gave me a lot of grief. I used an LM386 chip for the microphone amplifier. RF and AF combined in an NE-602 mixer. This idea came from British Columbia, from the "Wee-Willy" 80 meter transceiver design of Dick Pattinson, VE7GC. At one point I was having trouble getting sufficient carrier suppression. I thought something was inherently deficient in the circuit design, so I decided to use the somewhat more complicated balanced modulator circuit employed in the popular Elecraft K-2 transceiver. Later I had gone back to the simpler Wee-Willy approach (it works very well).

The variable crystal oscillator that is really the heart of this rig also underwent significant revisions over the years. As I said, it started out as a crystal oscillator on 17 meters. In an effort to get coverage of more of the band, I used two different crystals; they were mounted on the terminals of a big double pole switch on the front panel. Later this switch held 20 meter crystals. Later still, I discovered ceramic resonators. This was yet another idea from SPRAT. Ceramic resonators are cheap little devices that have most of the stability of quartz crystals, yet can be "pulled" in frequency much further than quartz. My Variable Crystal Oscillator became a Variable Ceramic Oscillator. Mike, KL7R, had suggested that I try putting two ceramic resonators in parallel in my circuit. I tried this, and discovered that instead of just making the circuit more stable, adding the second crystal somehow shifted the frequency range of the device by about 30 kHz. Perfect! With one resonator, it tuned from 14.274 to 14.305 MHz, and with a second resonator switched in, it tuned from 14.301 to 14.330 MHz. Now I could cover much more of the band. I may have been the first person to use ceramic resonators this particular way. This little discovery became my first contribution to SPRAT:

A Variable Ceramic Oscillator for 20m Phone
Bill Meara, M0HBR, N2CQR, CU2JL
meara.london@virgin.net   http://www.qsl.net/n2cqr

I'd been in search of more wiggle room for the VXO in my HB 20 meter DSB transceiver (see web site for pictures). Graham Firth and the G-QRP Club Store provided just the thing: Ceramic resonators at 14.3 Mhz.

Ian Macpherson's article in Sprat 73 provided the inspiration for the circuit. With just one resonator in the circuit, and using a 100 pf air variable I got a pleasing 31 kHz (14274-14305) of variation across a very useful portion of the 20 meter phone band.

During one of our morning (UTC) chats on ECHOLINK Mike, KL7R, suggested that I try two ceramics in parallel. I was very pleased to find that this shifted the entire frequency range up about 30 kHz (14301-14330). Wow! With a simple switch and two 50 pence resonators I was getting 56 kHz of coverage.

I quickly took the crystal oscillator out of the rig and replaced it with the "two ceramics and a switch" circuit. With a 180 pf air variable I got about 58 kHz of variation, with more overlap between the two ranges, so think the 100 pf air variable might be optimal.

You might have to play with the values of the feedback caps to get it running cleanly: on the prototype I used 330 pf caps, but in the rig I had to reduce the cap going to the base to 100 pf.

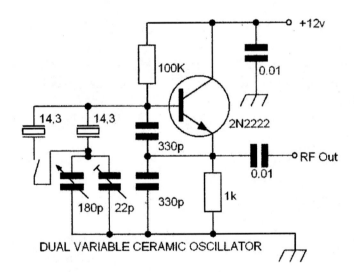

The RF amplifier in this rig also suffered through many bouts of revision, modification and "improvement." As you'd expect, taking an amplifier designed for the 40 meter band and using it on higher frequencies might cause trouble. It did. Convincing these amplifiers not to be oscillators was not easy. Eventually I scrapped the 40 meter design, and put in two amplifier circuits borrowed from some other source. Then I discovered that these two stages were not providing enough gain. But I had run out of space on the board, so I built a third RF amplifier stage and placed it up on the wall of the homebrew copper-clad box that housed this rig. (So it really was "driving me up the wall.")

Then I started reading about the many benefits of feedback amplifiers....

UNDERSTANDING: FEEDBACK IN AMPLIFIERS

Having fought so long and hard to avoid the feedback that turned my amplifiers into oscillators, you can imagine my puzzlement when I encountered amplifier circuits that deliberately included components that sent some of the output right back to the input. We are talking about negative feedback here, feedback that is out of phase with the input signal, feedback that deliberately reduces stage gain. Why would they do that? I started using these circuits before I understood the principles behind them. I was lured in by the promise of "unconditional stability." Yes sir! That's what I needed: unconditional stability.

That third amplifier stage that was built up on the wall of my DSB transceiver had feedback circuits in it. But I didn't understand how they worked, so I had to hit the books.

One of the reasons that feedback is deliberately introduced in amplifier circuits is to give the designer better control over the amount of gain. There can be big variations in the gain characteristics among transistors of the same type. This could have disastrous consequences for circuits intended to be reproduced; something that works in the designer's prototype might not work with the transistors used by others, even if these transistors are of the same type. Feedback circuits allow the stage gain to be determined not by the characteristics of the individual transistor, but instead by the amount of feedback.

But why? Why would negative feedback render irrelevant even large variations in the gain characteristics of individual tubes and transistors? As is often the case, I had to go through several technical tomes before I found a book that really answered my "how does it really work" kind of questions. Enlightenment on negative feedback came to me from the days before Pearl Harbor, from War Department Technical Manual TM-11-455 Radio Fundamentals, published on July 17, 1941. As we go through this, keep in mind that at times we will be talking about the overall gain of the entire stage (the transistor and the associated parts), while at other times we will be talking only about the gain characteristics of the particular transistor itself.

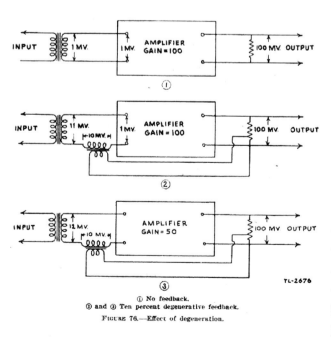

① No feedback.
② and ③ Ten percent degenerative feedback.
FIGURE 76.—Effect of degeneration.

Figure 1 shows an amplifier that increases the input voltage by a factor of 100. In Figure 2, 10% of the output voltage is fed back – out of phase – to the input. Now, when trying to understand feedback amps, you can find yourself trying to figure out how changes to the input produce changes to the output, which in turn produce changes at the input, etc. Like the amplifier you are trying to understand, you become like a dog chasing his tail. TM 11-455 elegantly breaks this cycle. It has us think this way: assuming we maintain 100 mV at the output, with 10% fed back, how much of an input signal will we need? Well, the amplifier device (say a tube or transistor) itself still amplifies voltage by 100. So we still need a net voltage of 1 mV at the input. So now we need a signal voltage of 11 mV to produce the 100 mV output (Figure 2). The feedback cancels 10 mV of the 11 mV coming into the stage, putting 1 mV on the input. Negative feedback reduces gain.

But how does negative feedback reduce the impact of variations in the gain characteristics of the individual devices? Without feedback, if we substituted our 100X amplifier with a device that happened to have a gain of only 50, our 1 mV input would produce only 50 mV out, possibly playing havoc with our overall design. We'd need an input of 2 mV -- a doubling of the input voltage -- to make up for the variation, to get our desired 100 mV output. But look at the situation with feedback (Figure 3). We still assume 100 mV at the output, and 10 percent negative feedback. But now, even though device gain has dropped from 100 to 50, we would only need an increase in signal input from 11 mV to 12mV. With this feedback arrangement, if we kept the input signal level at 11mV, and then used a transistor that happened to have only HALF the voltage gain of the original device, the output voltage would drop only to around 92 mV, a drop of about 9% -- not the 50% drop we saw in the circuit without feedback. Negative feedback reduces the impact of gain variations from device to device. It makes our amplifier stages more "beta independent."

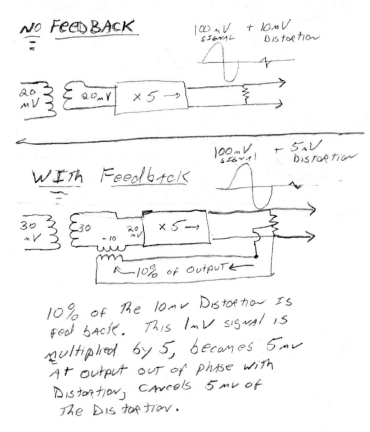

The textbooks also tell you that negative feedback reduces distortion. This statement is usually followed by a number of relatively simple equations that demonstrate mathematically how distortion is reduced, but as always, I was left scratching my head, wondering HOW this really works. Here's how:

Following Fourier's advice, let's think of distortion as an additional waveform riding along with our desired signal. In the diagram we have a 5X voltage amplifier with 20 mV at the input, let's say that it produces a complex distorted waveform that consists of our desired 100 mV sine wave, along with an ugly 10 mV distortion signal.

The feedback network takes 10% of both signals and feeds them back to the input (with a 180 degree phase shift). At the input, for the desired signal, the 10 mV of feedback meets up with 30 mV of input signal (as in TM 11-455, I'll keep output the same, but will account for the loss of gain by increasing the input); we end up with 20mV at the input to the amplifier device. This then goes through the 5X amp and we get our 100 mV output.

But look what happens to that ugly distortion signal: it arises IN the device. When the feedback portion of this distortion gets to the input, it does NOT meet up with an input signal. It just goes back through the amp. So the feedback network takes 10% of the 10 mV distortion, introduces a 180 degree phase shift, and sends this 1 mV waveform through the 5X amp. At the output of the amp we can think of the original 10 mV of distortion combining with what is now a 5mV out-of-phase signal. In this case, half of the distortion signal is canceled. We can say that compared with the no-feedback amplifier, distortion has been reduced from 10% to 5%. This circuit discriminates against distortion signals that arise inside the device. The feed back of the desired signal meets up with the input signal, cancels a portion of it, but then the remaining signal goes through the amp, producing the desired amplified signal. But the feedback from the distortion signal has nothing to meet at the input. It just goes through the amp and then cancels a portion of distortion signal at the output. More desired signal, less distortion.

Another reason to use feedback is to limit the gain to certain frequency ranges. Transistors inherently have more gain at lower frequencies. This can be one source of the dreaded "amplifier-turned-oscillator" problem. One way to combat this is to set up a feedback network that results in more negative feedback at lower frequencies. In this way your amplifier will work well at the operating frequency, while any possible oscillations at much lower frequencies would be quenched by the larger amounts of negative feedback.

One of the most useful reasons for introducing feedback is to get better control over the input impedance of an amplifier stage. From my days in the electronic kindergarten, when I was just slapping together bits of circuits, I learned the hard way that you can't just take the "goes-in-to" of one amplifier stage and connect it to the "goes out of" of the previous stage. Impedances have to be matched.

The ARRL handbook points out that the two main types of feedback are "degenerative" (essentially putting a resistor in between the emitter and ground on a common emitter amplifier) and "shunt" (shunting some of the actual output signal back into the input circuit.) (In some books ALL forms of negative feedback are referred to as "degenerative.")

I could easily see how degenerative feedback limited stage gain, but I had trouble understanding how shunt feedback affected input impedance. The handbooks pointed out that additional resistance in the emitter circuit increases input impedance. That's very intuitive—the signal at the input of the stage goes through the base to the emitter to ground. More resistance from emitter to ground should mean greater input impedance. So far so good. But then the handbooks usually point out that "shunt" feedback reduces input impedance. I had a tough time understanding that.

As often happens, enlightenment came via notebook and pencil. I just drew a little diagram of the relevant portion of a standard common emitter amplifier. Then I started thinking about what happens at various points in that amplifier as an input signal goes through its cycle. It is important to remember that this kind of amplifier is "inverting", i.e., when the input signal is going positive, the output signal will be going negative.

The feedback circuit creates a connection between the input at the base of the transistor, and the output at the collector. Without that feedback connection, when the input signal goes positive, it will "pull" a certain amount of current up through the transistor, through the emitter circuitry. How much current it pulls is determined by the transistor's characteristics, and by whatever other components are in the circuit (like that degenerative resistor you put between the emitter and ground). That relationship between the input voltage and input current defines the input impedance.

Now let's add that shunt feedback circuit. Now there is an additional path for current. With this shunt circuit, when that signal at the input goes positive, the input circuitry is connected to the now strongly negative collector terminal. Some of those electrons at the collector will flow through the feedback circuit into the input port. So for the same input voltage, the introduction of the feedback circuit has resulted in more current flow in the input circuitry. That's the definition of "lower impedance."

With feedback, the circuit designer has some very useful tools. He can control the gain of the stage. He can fight the tendency of the transistor amplifier to oscillate at low frequencies. And he can choose combinations of degenerative and shunt feedback that will give him the input impedance that he wants. Very often 50 ohms is the desired impedance.

****

Now that I understood this wonderful design tool I wanted to make more use of it. I had only very vague notions of the input impedances and gains of the various bits of circuitry that I'd thrown together. But now, with nothing more than a few resistors and capacitors, I had the ability to put those parameters right where I wanted them. I quickly turned the other two RF stages in my 20 meter DSB rig into feedback amps.

Slowly but surely my reform and restoration of the DSB transceiver was moving forward. Each evening at 8:30, I'd join the guys on 14.320 MHz and ask them how the rig sounded. At the start of the process, I could tell that they were having trouble hearing me. On other nights, I got the sense that they could hear me, but—because of the ugliness of my signal—kind of wished they couldn't. But they were all very nice about it (often too nice) and I think they got a kick out of helping me get this little rig into shape. Reports from Giorgio, I0YR, however, were notoriously unreliable. I think Giorgio is simply too nice to ever give a fellow ham a bad signal report. My signal could have sounded like a chain saw modulated by a jack hammer, and Giorgio would have told me "Tutto bene Bill! Cinque per nove! Tutto a posto!" ("All is good Bill. 5 by 9. All OK!")

It is usually difficult to make friends in a foreign city. There are often cultural and linguistic barriers, and most people are busy with already established work schedules and social routines. But Rome has to be one of the easiest places in the world for an outsider to make friends. Romans are very friendly, and very open to outsiders. They are "people people." Friendship is important to them. You can see this in the words that are used when friends converse. Not long after the beginning of our 8:30 meetings, I was being greeted as "Carissimo Bill" That roughly translates as "Very dear Bill." Whenever we'd wrap things up at 9 or 9:15, each participant would say goodnight to the "cari amici" ("the dear friends."). (A British author warned an Italian friend who was preparing to go to England to play professional soccer to drop all the "dears and dearests" when talking to his English teammates. The Italian friend seemed genuinely puzzled by this advice—for him, expressing affection for friends was completely normal.)

Of course, ham radio made it much easier; the welcoming culture of "CQ" combined with the friendliness of "carissimo" make Rome a great place to be a radio amateur.

CHAPTER 8
CONCLUSIONS
A BROTHERHOOD WITHOUT BORDERS

My home-brew projects started before I even knew amateur radio. I grownup with the school's boy desire to do something marvelous. I started building up my project from school's physic books . I thought myself a lot of basic skills step by step by using what ever hardware available at my home . Even a simple thing like the difference between a bare wire and an enamel wire i discovered after a several trials of winding up an electromagnetic coil. No one helped me except my family admire . At that time i was so eager to know more and more a bout the facts and rules of physic and electricity. And the real problem facing me is the lack of books and magazines !! Oh how much i suffered from that !!. Still i can remember when i fixed up a morse telegraph between two rooms at my home . And the first spark transmitter i built. Now i can listen and talk to the world with my home made radios

—Dr. Nader Abd Elhamed Ali Omer, ST2NH, Khartoum, Sudan

I guess it is inevitable: If you have a hobby like ham radio, a hobby that lets you reach out across borders… if you spend decades in far-off foreign places, staring up at distant galaxies and making friends with local people who share your interests in the electronic mysteries of the universe, well, over time the lines that we have scribbled on our maps and globes will start to lose their significance. For me, there is a lot of irony in this—it was, after all, ham radio that got me interested in international life, and that caused me to move into diplomacy, a career that at times seems all about emphasizing the importance of those little lines, the importance of our national differences.

Living overseas can sometimes make those little lines seem even more important. If you live in some form of enclave—a military base or a diplomatic compound—and have little contact with the people who live "outside the wire," that barbed wire becomes a very important local equivalent of those lines on the maps, those borders. Inside is US. Outside is THEM. And if you don't speak the local language, that wire takes on an added importance.

But if you really live in a foreign society, if you live among the people of the country and learn to speak their language fluently, then over time the national distinctions that once seemed so important seem to kind of fade away. Sending your children to the local school will greatly accelerate this blurring of national distinctions. You will have daily reminders that the common concerns of parents transcend borders. Send your kids to an international school, to a school where there are passports of fifteen or more different colors in each classroom, and the fading will accelerate even further. If you ask your kids about little Asaaf's citizenship, they probably won't even know. They don't care. It doesn't really matter.

Being a ham while living abroad seems to increase this fading of boundaries, this blurring of national distinctions. The hobby gives you important common ground, indeed a common obsession, with people who the folks back home would see as completely foreign, completely alien. My first peek at this world without boundaries came through articles in QST magazine about those intrepid foreign hams and their exotic homebrew gear. The stations that I put together as a teenager allowed me to communicate directly with them. Then my travels began, and in just about every place I've been, I've found guys whose lifelong interest in radio and electronics is eerily like my own, guys like Hilmar, Pericles, Gustavo, Messias, Giorgio, Gianfranco, Walter, and Fabio, guys who somehow got interested in this stuff in their early teens and who have remained interested in it throughout their lives. The fact that none of my American work colleagues would have understood my interest in radio (indeed, they might have ridiculed it) strengthened this sense of identification with my brother hams in the local community.

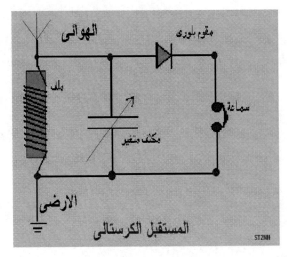

The SolderSmoke podcast intensified this sense of cross-border identification. Our programs have listeners around the world. When Mike was killed, the expressions of grief rolled in from all of corners of the globe. Friends in India needed time to regain their composure when they heard what happened. Messages came in from Australia and New Zealand, from Lebanon, from everywhere. The solder smoke blows across the borders.

"The Knack"—Dilbert's disease—is what we have in common. In the Steven Spielberg movie "Close Encounters of the Third Kind," people all over the world suddenly—inexplicably—have the urge to build models of one particular mountain. Something similar has happened to us—we have the urge to build shacks, and in them, radios.

In the years since Marconi first started tapping out his signals, a wonderful set of informal rules, principles, and values—a distinct ham radio culture—has risen up in the solder smoke. This is the culture of CQ, the culture of the Elmer, the culture of mutual assistance. This culture binds us together; we all struggle with the symptoms of the Knack, but we don't have to struggle alone.

Of course, this is just a hobby, and it is all for fun, so there are no strict parameters on what we do. But I think the internet has made us more aware of the fact that we Knack victims, we melters of solder, do share some common traits: We'd rather build than buy. We'd rather repair than replace. We like to understand the circuits and devices we use. We appreciate clever and economical solutions to technical problems. We enjoy the brotherhood of like-minded radio fiends. We have shacks.

We have a precious tradition of mutual support that goes back to the earliest days of radio; we help each other, and our tradition of Elmer-ism means that we are especially kind to newcomers. Our junk boxes are not seen as precious, private, personal stashes, but rather as elements of a global, communal junk box—when we throw some weird item in our collection, we don't necessarily think, "*I* might need that someday," but instead "*somebody* might need that someday."

We build things for fun, and we don't really care if they look weird or are weird. We can put microprocessors in the same rig with 6L6s. Ugly is OK with us.

We put soul in our new machines. They become magic carpets that carry our voices across oceans and borders, and into the shacks of unmet friends.

This is obviously the kind of book that lends itself to updates and revision. Having been born in the International Geophysical Year, with any luck I may be able to operate through two or three more solar cycles. I'm still in the Foreign Service, so even though it looks like Washington will be our next stop, there is the possibility of more foreign adventure, the possibility of more chapters for this book. There are still a lot of things I haven't tried yet in ham radio: there's still moon-bounce, meteor scatter, 160 meters, amateur television, and many other things. And of course—as readers will have certainly noticed—there are many things about radio that I still don't understand: I'm really shaky on the Smith Chart and the design of narrow-band amplifiers... I think I understand the waves that are emitted by my antenna, but I have a lot of trouble thinking about 20 meter photons... I'm still not quite sure how my SWR meter separates the forward and reflected signals... Maybe we can explore these and other mysteries in future editions of this book... Please stay tuned!

Acknowledgments

As I look around my Rome radio shack, I see many reminders of who helped make this book possible:

Right behind me is the workbench that supported many of the projects described in these pages. It was a Father's Day gift from my wife Elisa, and I think it beautifully captures the great support that she has always given to my eccentric electronic endeavors.

I also see many things belonging to my children, Billy and Maria. They are a constant source of inspiration and wonder. This book is dedicated to them.

Some of the items that surround me come from the days of my electromagnetic youth in Congers, New York. They remind me of the great support that I always received from my mother and father. Every time I buy a radio book, I think of my mom diligently saving coupons to buy us a new encyclopedia, or saving pennies to get me C.L. Strong's "The Amateur Scientist." And every time I walk into a radio club, I think of my dad driving me up to the Crystal Radio Club in Valley Cottage, N.Y., and waiting patiently for the club meeting to finish.

The computer and the radios that sit in front of me, remind me of all the inspiration, friendship and fun that have come my way from my fellow radio amateurs from around the world. Whether the contact was made by ancient Morse code, or via a satellite, or over the internet, these contacts provided the motivation to build circuits and to try to understand them.

The bookshelf in the shack reminds me of the debt that all of us have to the American Radio Relay League. A good portion of those books were published by the League. I thank the ARRL for publishing them, and for kindly allowing me to use materials from them, and from QST, in this book.

Also on those shelves is my cherished collection of SPRAT magazine. SPRAT is the quarterly journal of the G-QRP club. It has been a key source of inspiration and ideas over the years. We all owe a debt of gratitude to George Dobbs, G3RJV, and the entire SPRAT team.

George Katzenberger, K8VU, volunteered to edit this book, and diligently worked on it even during an extremely busy time for him. Todd Kanning, K5TAK; Jack Welch, AI4SV; Steve Hartley, G0FUW; and Andrew Atkinson, G4CWX also went through the book in search of my mistakes. They all caught many errors and provided great advice. The book is lot better for their efforts. Of course, I'm completely responsible for any that slipped past them, and for everything else in the book.

Please Help Improve This Book!

This is very much a "homebrew" book. I wrote it on a fairly ancient Dell Computer running (gasp) Windows 2000. I published it myself using the Lulu on-demand-printing service. My diagrams are hand-drawn --- I admit to being software challenged in this area.

What I'm saying here is that this book could use some help. If you spot errors please let me know and I'll correct them for future editions. If anyone out there wants to spruce up the diagrams, please do! Send them to me and I'll put them in the next edition (with due credit to the draftsman). And any other suggestions for improvement will be gratefully received.

Thanks a lot!
Bill Meara
bill.meara@gmail.com

INDEX

Abbott, Ian, 167
Adams, Scott, 5
Ali Omer, Nader, 191
Amado, 51, 52, 56, 80
amplifier, 12, 33, 34, 36, 43, 50, 56, 63, 86, 88, 89, 102, 103, 106, 107, 112, 121, 122, 123, 124, 125, 129, 130, 132, 136, 139, 140, 143, 145, 146, 147, 148, 149, 157, 160, 168, 169, 170, 171, 175, 177, 182, 185, 186, 187, 188, 189, 190
amplifier loads, 1, 20, 43, 55, 122, 170, 171, 172, 173
Angola, 3, 177
AO-21, 53
Apt, Jay, 77, 78, 80
Arianrhod, Robyn, 14
Armstrong, Edwin Howard, 90, 91, 94, 96, 97
ARRL (American Radio Relay League), 13, 14, 17, 18, 27, 31, 53, 71, 97, 104, 108, 110, 113, 131, 169, 189, 194
Asimov, Isaac, 8, 34, 62, 74, 133
astronomy, 3, 6, 52, 56, 57, 79, 139, 143, 161
balanced modulator, 107, 122, 123, 124, 125, 128, 129, 132, 133, 136, 146, 147, 148, 160, 185
Barbosa, Jorge, 3
Barebones Superhet, 85, 101, 104, 105, 121, 147, 166
Bettencourt Faria, Carlos Mar., 3
Bingham, Dick, 130
BITX 20, 116, 168
boatanchors, 85
Brannas, Chris, 139
Britton, Lindsay, 139
Bryson, Bill, 116
Burrell, Jim, 177
capacitor, 33, 60, 61, 62, 64, 65, 66, 68, 69, 70, 71, 89, 104, 111, 120, 125, 133, 135, 150, 152, 158, 160, 172, 177
Castelnuovo, Giorgio, 181
Caughan, Mike, 166
Citizens Band, 12, 26
CQ, 2, 23, 24, 25, 63, 72, 73, 126, 127, 128, 129, 130, 137, 139, 149, 150, 157, 165, 185, 190, 192
Crystal Radio Club, 13, 17, 35, 36, 47, 82, 119, 182, 194
crystal radios, 89, 167
Dell, 194
DeMaw, Doug, 85, 91, 92, 101, 103, 104, 121, 122, 128, 129, 130, 132, 133, 136, 168, 177, 185
Dennison, Bob, 66
DeSoto, Clinton B., 1, 81
Dilbert, 5, 7, 192
diodes, 36, 42, 43, 86, 87, 89, 90, 95, 96, 105, 123, 125, 128, 129, 130, 132, 135, 136, 148, 159, 160, 167, 177
dipole antenna, 18, 19, 27, 51, 52, 82, 117, 125, 144, 145, 158, 163
direct conversion receiver, 36, 66, 91, 92, 121, 125, 128, 130, 131, 145, 146, 185

Double Sideband, 105, 107, 108, 121, 122, 123, 124, 125, 126, 127, 128, 129, 130, 133, 146, 148, 149, 157, 159, 160, 161, 184, 185, 187, 190
Drake 2-B, 1, 29, 35, 43, 44, 45, 46, 47, 53, 54, 55, 56, 59, 63, 72, 82, 91, 103, 110, 118, 119, 121, 122, 123, 126, 128, 132
DX-100, 12, 18, 25, 27, 30, 83, 84, 109
Echolink, 140, 164, 165, 166, 167
Edison, Thomas, 180
Einstein, Albert, 14, 32, 33, 67, 68, 132, 160
electrons, 7
Elmer, 13, 17, 192
Elser, Fred Johnson, 5, 53
Experimental Methods in RF Design, 99, 176, 184
Faraday, Michael, 14
Farhan, Ashhaar, 116, 168
feedback, 63, 103, 122, 132, 139, 152, 166, 167, 176, 184, 186, 187, 188, 189, 190
Fessenden, Reginald, 11, 90, 150
Feynman, Richard, 7, 14, 62, 153, 171
Filippi, Walter, 183
Firth, Graham, 178
Fishpool, Tony, 178
Fourier, Jean Baptiste, 99, 100, 101, 108, 133, 135, 137, 188
Galeffi, Fabio, 180
Garriott. Owen, 74
Globe Electronics "VFO Deluxe, 18
G-QRP Club, 66, 91, 146, 147, 161, 194
Grammer, George, 60
Green, Wayne, 75
Gunn, James E., 180
Hallicrafters HT-37, 33, 35, 37, 43, 45, 46, 47, 50, 54, 55, 56, 59, 72, 82, 105, 106, 107, 108, 109, 110, 126, 128
hamfest, 60, 64, 83, 84, 85, 108, 110, 111, 113, 140, 147, 160, 178
Hammarlund, 83, 109, 110
Hayward, Wes, 130, 168, 169, 177
Heathkit, 12, 18, 31, 35, 83, 84, 85, 110, 119, 120, 147, 166, 181, 182, 183
Hertz, Heinrich, 16, 25
Holden, Tom, 95
Hopkins, Michael, 113, 139
Hopkins. Michael, 104
Hubble Space Telescope, 58
Hudson, Jack, 136
Huff, Paul, 183
International Space Station, 140
Ippolito, Piero, 181
James, Tom, 72
Jensen, Peter, 15
Jones, Frank, 104, 113
Jupiter, 6, 16, 56, 57, 58, 59, 143, 161
Jurgens, Tom, 63
Katzenberger, George, 194

195

Kidder, Tracy, 178
Kites, 154
Knoll, Ed, 63
Lafayette HA-600A, 10, 11, 12, 29, 84
Larsen, David, 79
Larson, Erik, 15, 46
lasers, 42, 86, 167
Lasses, Rolf, 140
Laun, Fred, 48, 49
Lee, Frank, 146, 150, 152
LTSpice, 174, 175, 176, 185
Luecke, Jerry, 136
Maier, Hilmar, 13
Manhattan College, 43
Marconi, Guglielmo, 11, 15, 16, 25, 27, 46, 63, 150, 192
Masters, Percy, 138
Mathes, Stanley, 5, 53
Maxim, Hiram Percy, 53
Maxwell, James Clerk, 11, 15, 16, 23, 33, 34, 40, 68, 69, 104, 132, 160
McCoy, Lew, 110, 112
Meara, Billy, 5, 68, 114, 116, 117, 118, 120, 141, 167, 177, 178, 183, 194
Meara, Elisa, 45, 74, 79, 82, 85, 101, 139, 140, 153, 155, 165, 181, 194
Meara, Maria, 5, 114, 116, 118, 161, 162, 178, 183, 194
Michigan Mighty Mite, 46, 63, 64, 66, 71
Mighty Midget transmitter, 110, 111, 112, 113, 114, 147, 181
mixers, 36, 91, 92, 93, 94, 95, 96, 97, 99, 101, 108, 112, 128, 129, 130, 131, 132, 135, 136, 147, 148, 150, 151, 152, 159, 160, 185
modulation, 105, 106, 107, 133, 135, 137
Moniz, Messias, 119
Morse code, 3, 13, 16, 17, 18, 23, 24, 25, 26, 29, 30, 35, 43, 44, 45, 49, 54, 55, 56, 66, 71, 73, 92, 103, 104, 137, 143, 155, 178, 182, 194
Mottley, Ray, 142
Nahin, Paul, 14
Newell, Maurice, 137
Newkirk, Rod, 27
Norgaard, Donald E., 108, 109
NOVA-QRP Club, 114
novice license, 13, 14, 16, 17, 19, 26, 30, 31, 110
oscillation, 63, 64, 104
P3D, 118, 140
Parfitt, Dale, 105
PCsat, 140
Perdomo, Pericles, 49, 50, 54, 55, 75, 80, 104, 111, 147, 150, 192
Power Formula, 20
Preston, Richard, 180
quartz, 18, 65, 159, 185
Radio Club Dominicano, 48, 80
radio shack, 1, 9, 12, 13, 18, 27, 28, 31, 35, 45, 46, 47, 49, 50, 55, 73, 76, 77, 81, 82, 86, 114, 116, 117, 118, 119, 137, 141, 143, 145, 152, 153, 155, 157, 158, 159, 162, 164, 166, 170, 178, 181, 183, 184, 194
resonant circuit, 63, 68, 170
Rocketry, 52, 153, 154
Rockey, C.F., 149
RS-10, 53, 55, 56, 74, 118, 143
RS-12, 1, 2, 54, 55, 56, 118
satellites, 3, 52, 53, 54, 55, 56, 58, 75, 80, 118, 140, 143, 162
Scasciafratti, Gianfranco, 181
Schick, Rolf, 138
semiconductors, 37, 38, 89
Shepherd, Jean, 9, 12, 27, 83, 165, 166
Shoemaker-Levy 9, 57
Signal School, 40, 41
Silva, Francisco, 144, 145, 153
Smith, Brad, 167
Solid State Design for the Radio Amateur, 130, 168, 184
SPRAT, 97, 146, 150, 154, 185, 194
Stokes, Ed, 139
Stoll, Clifford, 45
Strong, C.L., 6, 194
superheterodyne, 91, 96, 146
Swan 240, 104, 111, 147
Tait, John, 139
Tartini, Giuseppe, 94, 95
Taylor, Joe, 157
Taylor, Jonathan, 164
Thagard, Norm, 74, 76, 77, 79, 82, 143, 165
The Knack, 5, 7, 108, 181, 192
Vasquez, Gustavo, 48
Vienna Wireless Society, 82
Voice-over-Internet, 73
W1AW, 13, 17, 26, 104
Weiss, Adrian, 28, 37
Welch, Roy, 53
Williams, Jim, 183
Wilson, F.A., 8
Wozniak, Steve, 5, 167
Yaesu Memorizer, 43, 55, 56, 143, 165

ABOUT THE AUTHOR

The author in 1994

Bill Meara has been melting solder and throwing antenna wires up into trees since he was twelve years-old. Because of his day job (diplomat), he has lived and worked in Central America, the Dominican Republic, Spain, Portugal (The Azores), the United Kingdom, and, most recently, Italy.

He is the host of the SolderSmoke podcast (www.soldersmoke.com), and he blogs at SolderSmoke Daily News (soldersmoke.blogspot.com). He has published numerous articles in amateur radio magazines, and is the author of the non-fiction book *"Contra Cross"* which was published in 2006 by Naval Institute Press.

He lives in Rome with his wife, Elisa, and their two children, Billy and Maria. His e-mail address is bill.meara@gmail.com

Also by Bill Meara…

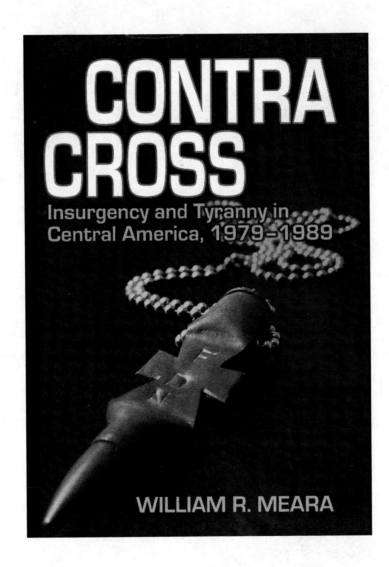

Available in paperback, as a download, or as a Kindle e-book:

http://contracross.com